Next

The Coming Era in Science

Next

The Coming Era in Science

Holcomb B. Noble

General Editor

Little, Brown and Company   Boston   Toronto

FIRST EDITION

Library of Congress Cataloging-in-Publication Data

Next, the coming era in science.

1. Science. 2. Technology. I. Noble, Holcomb B.
II. Title: Coming era in science.
Q172.N46 1987. 500 87-22835
ISBN 0-316-61130-1
ISBN 0-316-61133-6 (pbk.)

FG

Published simultaneously in Canada
by Little, Brown & Company (Canada) Limited

PRINTED IN THE UNITED STATES OF AMERICA

Contents

Introduction   **WHAT'S NEXT?**   ix
                  Holcomb B. Noble

Chapter One   **THE BIZARRE AND SERENDIPITOUS**   3
                  HISTORY OF DISCOVERY
                  Leon M. Lederman

Chapter Two   **TOMORROW'S THINKING COMPUTER**   42
                  Richard Flaste

Chapter Three   **THE RACE FOR SUPERCOMPUTERS**   68
                  Peter H. Lewis

Chapter Four   **THE BIOCHIP:**   82
                  MERGING COMPUTERS AND BIOLOGY
                  Michael Edelhart

Chapter Five   **COMMUNICATIONS:**   95
                  AROUND THE WORLD IN SPLIT SECONDS
                  Andrew Pollack

Chapter Six   **THE EARTH**   131
                  AND ITS NEW SYSTEM OF SATELLITES
                  Wayne Biddle

Chapter Seven   **THE EXPLORATION OF SPACE**   142
                  John Noble Wilford

Chapter Eight   **GRAND UNIFICATION THEORIES:**   160
                  FAITH IN ULTIMATE SIMPLICITY
                  Timothy Ferris

Chapter Nine   **BEYOND THE KNOWABLE:**   173
                  THE ULTIMATE EXPLORATION
                  Holcomb B. Noble

Acknowledgement

Grateful acknowledgement is made to Cory Dean, assistant science editor of *The New York Times*, and Brenda Nicholson, administrative assistant in the newspaper's Science Department, without whose editorial skill and assistance the "Next" series could never have been completed. Similar gratitude is extended to Sam Summerlin, president of The New York Times Syndication Sales Corporation, whose energy and enthusiasm made it all happen.

Introduction

What's Next?

By Holcomb B. Noble

Holcomb B. Noble is deputy director of science news, The New York Times.

As the curtain begins to rise on the twenty-first century, it will start the process of revealing a world of science and technology never really dreamed of, never properly imagined even in extended realms of phantasmagoria. This is not a prologue to science fiction. This is not sensationalism or campfire storytelling or even highbrow futurism. This, for good or ill, is an eminently reasonable assurance. This, so far as one can make statements of fact about the spin of the earth, say, on the day after tomorrow, is a statement of fact. Its reason stems, if nothing else, from the experiences of the twentieth century as they stand in contrast to all that went before, and as they have set wheels in motion for events that are bound to follow. This book, broadly speaking, begins with a discussion of just that: all that went before as it is of interest in predicting what is to come; a history of the key events of science over the past 2,500 years. A key event in this sense is one that brings about significant change. The book's first chapter is by Dr.

Leon Max Lederman, a preeminent American physicist who is the director of the Fermi National Laboratory and has been awarded the American National Medal of Science. Dr. Lederman participated in the production of the first accelerator pion beam, which was used to study the properties and interactions of subatomic particles. But, more to the point, he is also possessed of that rare ability to combine complicated scientific and technical knowledge with a lucid, informed style of writing that is compelling and understandable. He writes in this volume for your grandmother, among others — that is, your persistent and highly intelligent grandmother, the one who, to her everlasting regret, has had little time to devote to science but cares deeply about her world, the one who is eager to know what lies ahead for her grandchildren. Dr. Lederman explains the mathematical genius of the Greeks and sweeps on through the centuries to Einstein's special and general relativity and Dirac's quantum theory.

The following chapters are by science writers on the staff of *The New York Times* or by free-lance science writers. Along with the theoretical, they deal with the practical application of scientific discovery and technological developments of the twentieth century as these happenings foreshadow what is to come. The computer revolution is, of course, one of the most practical and most dramatic. Moreover, computers will become vastly more powerful than they already have and vastly more intelligent. What will that power accomplish? For whom? To whom? Will the ultimate machine surpass human capability? Will it have a soul? With such technical wizardry being applied and imparted to it, why won't it become able to reproduce? What then? What will the consequences of the communications revolution ultimately be? Clearly, they will continue to be monumental, for the third-world countries as well as for those already more highly developed. The significance, moreover, will be every bit as great as earlier communications breakthroughs, as the printing press, the telegraph, the telephone, the simple transistor radio. The sending of written messages, that is, the sending of mail, at the speed of light, at 186,000 miles per second, will shrink the world, for better or worse — and it's hard to imagine that, on balance, it will be for worse. Communication has always been better than no communication, and, in communications at least, faster is better than slower. The inexpensive, portable, lightweight computer will eventually become as readily available, or, indeed, more readily available, than the telephone or automobile. It will have easy-to-understand functions and virtually all the memory storage wanted or needed. Instant communication, from city to city, office to office, campus to campus, home to home, will result. But will there be no harmful ramifications? What about invasion of privacy? In short, how will more of these communication breakthroughs come about and what will their implications be for industrial production, academic research, technological development, education, publishing?

How does one assess this twentieth-century trick of giving the planet Earth a whole battery of new artificial satellites? What is this all going to mean, ultimately? Putting spacecraft into geosynchronous orbit, 22,300 miles above Earth, for example, enables them to spin around Earth at a rate that corresponds to the spinning of Earth on its axis, and therefore, in effect, leaves them permanently fixed over one location. Use of such satellites will continue to expand greatly, particularly in terms of increased radio-television broadcasting facility, data and telephonic transmission, military reconnaissance, weather forecasting and interterrestrial observation. Crowding may eventually lead to large platforms that serve as giant switchboards in space. Besides geosynchronous satellites, there will be more intensive use of instruments in low Earth orbit for resource-surveying. Multinational space programs may encourage applications in this region, with possible commercial developments — such as materials processing. Still very much at issue is whether low Earth orbits could also become deadly new theaters of war, stagings for military space weaponry, a satellite that could fire X-ray lasers, for example, or particle beams at enemy satellites or approaching ballistic missiles. The spaceship *Pioneer*, which in the twentieth century became the first man-made machine ever to leave the planetary system, will spend much of the approaching decade traveling millions of miles among constellations unknown and radioing new astronomical data back to Earth. Among the goals of man for *Pioneer* and other manned or unmanned spaceships will be to obtain a fuller understanding of the origins of the solar system and the beginnings of the planet Earth itself.

This book will assess what the space programs of the various nations have accomplished, what in the months, years and decades to come will be undertaken and what of practical or theoretical value they will represent. Perhaps the most profound area of twentieth-century foreshadowing lies in the currently intensive research into and development of a Grand Unification Theory, the search for a single explanation for all the physical forces in the universe. Why would this make such an enormous difference in the twenty-first century? Einstein spent the last half of his life in search of it. Why was it so important to him? What the future holds once such a universal theory is evolved is perhaps as difficult to imagine as it would have been for the world to foresee the consequences of Maxwell's, Einstein's or Bohr's discoveries in the late 1800s and the early part of the 1900s — and perhaps equally charged with potential for change. Is the great unknown about to be known? Were the twentieth-century explosions in science and technology just the beginning? Will they pale in comparison with the twenty-first century? Probably so. The present always does. But will not tinkering with such fundamental concepts as space and time, as twentieth-century scientists began to do, alter the dimensions of the future in ways far more profound than ever before? It is these questions, these dimensions, known and unknown, that this book attempts to address.

Next

The Coming Era in Science

Chapter One

The Bizarre and Serendipitous History of Discovery

By Leon M. Lederman

Leon M. Lederman is the director of the Fermi National Laboratory of the United States.

Twentieth-century physics is the repository of two great intellectual revolutions and, with some uncertainty due to the lack of historical perspective, we are now in the midst of what may be a third equally profound alteration of our worldview.

Taken together, the quantum theory, the theory of relativity and the current frenetic activity in particle physics are elements of a coherent synthesis of the creation, evolution and structure of the physical universe. The earlier revolutions have brought the influence of physics to all other sciences as well as to philosophy and literature. The results have dramatically changed the way people live and think. This is a traditional role of science and is a repetition of the correspondingly dramatic advances of science in the seventeenth, eighteenth and nineteenth centuries. However, as the world shrinks because of modern communication, the impact of what has been and what may yet be tends to more rapid and more global amplification.

It is our purpose to review the major events of twentieth-century physics with some emphasis on current problems and current ideas of how it will evolve as we enter the next century. We found it irresistible to trace the threads back to origins in antiquity. The effort is to be concise, to stress the connective issues without clarifying details that would obscure the sweep of events. How else to go from Greek atoms to grand unification in fewer than 8,000 words?

Scientists have come to accept some pretty bizarre notions. The most precise theory ever created, quantum electrodynamics (QED), starts with one of its fundamental postulates being a statement of uncertainty. We deal with concepts of curved space, of a vacuum that is rich in physical properties, and of pointlike particles that have a radius equal to zero but which, without embarrassment, carry spin, electric charge, mass and a plethora of other endowments.

Furthermore, today's physicists, lacking any semblance of humility, assert that these concepts are leading us with considerable optimism toward the ultimate goal: a complete synthesis, in which a small number of objects, subject to a single, unified force, would encompass all the observations in all the laboratories of the world and would give a consistent account of the evolution of the universe from the instant of creation in the big bang to the present time and projecting to the infinite future.

In what follows it is important to distinguish clearly between science and technology or invention. Science deals with ideas and is a curiosity-driven, abstract, cultural activity. Technology deals with tools and other things that people use. Faraday's laws of electricity and magnetism are science. Marconi invented wireless technology. Clausius contributed to the science of thermodynamics. Watt invented the steam engine. It was science that clarified the nature of nuclear binding energy; but it was technology that, in an astonishingly short time, converted it to a weapon of unimaginable power. Whereas the first clear records of scientific concerns date back to about 600 B.C., the history of technology is much older. There is evidence that toolmaking goes back as far as one million years. Invention that produced technology did not require scientific reasoning until relatively modern times. The progress in technology was an important component of natural selection as our human ancestors learned to cope with recurrent ice ages and other conditions hardly conducive to creative contemplation. By the time of the apparently abrupt appearance of scientific thought, a considerably sophisticated technology was at hand: fire; metalworking; agriculture; weights and measures; elements of arithmetic, algebra and geometry; an astronomical data base; navigation; land surveying; medicine and surgery; calendars. It is a matter for detailed historical scholarship to determine how much of the technology that was available in the seventh century B.C. came about by curiosity and those drives that we now associate with pure research. If modern experience is any guide, a substantial component did come from the dreamers; it's hard to believe that stellar navigation, for example, was discovered by lost seamen desperately trying to find their way home on a dark night in the Aegean.

Twentieth-century technology is essentially all derived from the results of science. There is an intimate interweaving and mutual enhancement of these two disciplines that, in the past century, accounts for the ever-escalating pace of both: science begets technology, science uses technology to create more science. More science begets more technology.

This litany deserves illustration:

1. Science begets technology; for example, the quantum theory of solids leads to the transistor.

2. Science uses technology to create more science. The transistor provides fast, low-cost digital computers and electronic circuitry for data acquisition, controls and analysis. Electronic controls and computer guidance vastly improve the performance of electron accelerators that not only do research in particle physics but also produce synchrotron radiation.

3. More science begets more technology. The synchrotron X rays are used for lithographic etching of integrated circuits for more powerful application of transistors to the making of supercomputers.

In spite of the interdependency of science and technology, the two subjects are distinguishable. Physics itself is divided into "pure" and "applied" with a gradualness across the spectrum that permits further qualification: abstract and exotic, almost pure, not too applied, all the way to engineering "R&D." It is in the applied domain that there is the activity of making things, presumably useful and, in particular, providing improvements in the material quality of human life. But what appears to be abstract, remote from day-to-day concerns, changes with time. Today it is quarks and black holes; 150 years ago it was electricity.

The distinction between basic physics and applied physics is important for public appreciation of potential for social changes and for science policy and planning, but it should not be overstated. A research laboratory, whether devoted to the exotica of particle physics or to an investigation of new materials for space rockets, looks and smells pretty much the same. On the human level, the day-to-day problem solving, the exuberance of occasional insight or the discouragement of repetitive failures — these are the common ingredients of intellectual engagement. Modern science includes the profound, such as the search for the instability of the proton, and the routine, such as efforts to make a more precise measurement of the mass of the electron. Also, technological developments include not only research on the mechanism of organ rejection but also the development of digital toothbrushes. Given all of this, we must emphasize what has characterized the twentieth century: a clear recognition of the power of science, through technology, to have an ever accelerating influence on the conduct and fortunes of society.

To project into the future, the present is obviously not enough, and we must depend heavily on historical perspective. Before venturing into this domain, I use the disclaimer of the physicist Richard Feynman, who calls this kind of history "conventionalized myth-story" not to be confused with historical scholarship. This is an honorable task for a scientist. We tend to revel in recapitulation of the work of previous generations in order to "put in context" our own work. It is a way of stepping back to admire ourselves. But it is also required to demonstrate the cumulative and connected nature of scientific thought.

The words *cumulative* and *connected* require explication.

Observations and measurements provide data that accumulate from epoch to epoch. These form the stuff of hypotheses and theories that in turn suggest new observations. Ultimately we arrive at major syntheses. These also accumulate, building toward the goal of an ever more inclusive and simple conceptual structure. Connectivity involves the influence of one epoch, one school, one scholar on another, leading to a significant advance in the science. It is sad but true that historical relevance accrues not to the originator of an idea or a fact but to the person providing the connection. Credit for relevance goes not to the Greek genius Aristarchus of Samos, who first proposed, eighteen centuries before Copernicus, that the earth moves around the sun, since that insight was lost. But the connectivity of Copernicus to Galileo, Kepler and Newton was seminal to progress. It is precisely this connectivity which we will stress in this survey.

HISTORICAL OVERVIEW

For about 500,000 years or more, our hunter-gatherer ancestors roamed the planet with essentially the same genetic structure we have now. Somewhere in even more remote eternity, curiosity, a sense of wonder, humor — the ingredients of creativity — were injected into the evolutionary stream. It is only in the past 2,500 years that history records the appearance of scientific thought, that is, the replacement of mythological notions by attempts at rational models. Modern science connects to this transition as it took place in the Greek colonies. This began an evolution whose pace, with some notable setbacks, has continued to increase to this day with no sign of saturation.

One of the scholarly enigmas is the apparently simultaneous (give or take a few hundred years) explosion of intellectual activity in many different parts of the world: in China with Lao Tzu and Confucius; among the Hebrews with Ezekiel and Zechariah; and with the scientifically refreshing Buddha. Nevertheless, the connections lead us primarily to the Greek philosophers, starting in the seventh century B.C.

Greek science had a run of about 1,000 years during which the relevant issues were laid down. We are, today, dealing with these same issues and may be concerned with them for some time.

The charge to the coming generations of scientists epitomized by today's particle physicists and extragalactic astronomers was given most explicitly by the Milesian school starting about 650 B.C. in Ionia, an area colonized by the Greeks about 900 B.C. Miletus, one of the Ionian towns on the west coast of modern Turkey, washed by the Aegean sea, was a ferment of philosophical, literary and artistic creativity. There is ample experience to allow us to conclude that the transition from magic and ritual to mythology to science was in fact a gradual process. Science has, in fact, never been entirely free of myth and mystery; it appears again and again, even among the most productive of schools and individuals — from Pythagoras to Kepler to Newton. Nevertheless, a fairly sharp break is perceived in Miletus, in the seventh century B.C. Remember that this was the dawn of theoretical science; as such, the jewels of wisdom were obscured by layers of mystical encumbrances. However, when extracted and polished by later scholars, they provide the connections we seek. If science is characterized by systematic positive knowledge and an active search for synthesis, then the beginning was in Miletus. Here were created two broad, philosophical ideas: (1) The universe is subject to rational inquiry and is potentially explainable "by ordinary knowledge." (2) One should seek for fundamental, primordial constituents (elements) out of which the things of the world, in their infinite variety, are composed. This Greek emphasis on parsimony, that the ultimate description of the physical universe must be simple, has been a fruitful guide to physics up to present times.

Thales of Miletus proposed water as the basic element and perhaps deserves the credit for having authored the simplest of all systems. Greek atomic theory developed with Thales, Anaximander, Heraclitus and Anaximenes as they debated the basic philosophical issues of primordial constituents. Eventually Empedocles proposed the four elementary objects — air, earth, fire and water — and two opposing forces, which he called love and strife. This idea persisted far beyond the tenure of Greek science. The concept of force was vague but led to a description evocative of present concepts of fields (effluences). The description of atoms closest to our present knowledge was given by Leucippus and Democritus about 400 B.C.

Democritus displayed the cheerful optimism of modern particle physicists by noting that a universe composed only of atoms and the empty space in which they move cannot be too difficult to comprehend. Atoms are extremely small and indivisible. By their combinations they make all the things that are seen. A century later (about 300 B.C.), Epicurus and his followers embraced the atomic view but added their own confusing notions. In the origins of a philosophical debate that would wax with Newtonian mechanics and wane with the quantum theory, Epicureans used the indeterministic motion of atoms to support the notion of human free will. A few centuries later, the Roman poet Lucretius would sing of the theory of atoms with a clarity and prescience that continues to enthrall us moderns.

It is impossible to overstate the importance and sweep of the concept of atomism as developed by these philosophers. If, in the twenty-first century, books will have built-in audio systems, this is the place for a roll of trumpets. The power of the microscopic viewpoint deserves clarification by a few modern examples. Chemists knew that when hydrogen is burned, two atoms of hydrogen combine with one atom of oxygen to form a molecule of water, H_2O. However, in response to "why?," physicists have given a deeper explanation in terms of the nature of the force between electrons and nuclei and in terms of the energy levels and dynamics thereby induced. Having gained a command of the underlying mechanisms, one can go on to explain, to predict and indeed to design new molecules and new chemical products. Another simple example: a bar magnet is a familiar enough object with the curious property that the north and south poles cannot be separated. Cut the bar in half and we have two bar magnets. At the superficial level, much was written in the old physics texts on the laws of magnetism and the properties of bar magnets. But it was only when the electronic structure of the iron atom was understood in terms of orbiting and spinning electrons that a complete understanding of static magnetism was at hand. This also led to understanding of the magnetic properties of other atoms and to their ability to predict and indeed to design new magnetic materials. That physical processes must seek their fundamental explanations via microscopic mechanisms became a profound dogma of science and underlies the success of modern chemistry, biology, genetics, materials science, physics and even cosmology, as we will soon learn.

Greek science was a world populated only by theoretical physicists. Like today's cosmologists, they observed the world, *as it was*, for their data. This process of interpreting nature was carried forward by the vigorous and durable school founded by Pythagoras in about 540 B.C. This took place in another realm of the Greek empire, in southern Italy. Information about the guiding principles is sparse and indirect but usually associated with emphasis on mathematics and a fascination with harmony of celestial motions and the primacy of arithmetic integers ("number is everything"). One wonders about the effect when Pythagoras discovered the law of the right triangle and the fact that the length of the hypotenuse of a triangle with unit sides is the square root of two. They must have considered the result totally irrational! Some confirmation of faith in integers came 2,000 years later with the equations of the quantum theory and solutions in terms of quantum numbers.

The giants of this last Greek period were Euclid, Archimedes and, later, Ptolemy of Alexandria. The latter's astronomical textbook was on the best-seller list for 1,400 years or so, until replaced by the work of Tycho Brahe. Euclid's geometry textbook is still in use.

Archimedes, in the second century B.C., developed a style that would not arise again until Galileo. He measured, set forth hypotheses, worked out logical consequences and subjected these to experimental test. He founded the subject of mechanics, applying the deductive rigor of geometry to theories about levers, pulleys and, of course, hydrostatics (when he cried, "Eureka!"). His works were avidly sought by Renaissance scientists. The total body of Archimedes' work — philosophy, mathematics, astronomy — would remain as the most enduring monument to Greek civilization.

Greek science derived much from the technology of the

Babylonians and the Egyptians. Is there an "essence" to the Greek departure? The answer is given most concisely by W.K.C. Guthrie, a recent scholar of Greek science:

"The Greeks asked 'why?' and this interest in causes leads immediately to a further demand: the demand for generalization. The Egyptian knows that fire is a useful tool. It will make his bricks hard, will warm his house, turn sand into glass, temper steel and extract metals from ores. He does these things and is content to enjoy the results. But the Greeks asked *why* the same thing, fire, does all these different things. . . . Then one is asking, what is the nature of fire in general, what are its properties. This advance to higher generalization constitutes the essence of the new step taken by the Greeks."

The "higher generalization" would eventually lead, centuries later, to a total comprehension of the process of fire and its mastery for a new level of technology that would lift humanity to undreamed-of levels of fulfillment.

The quality of mind that enabled this unique society to flourish in science must have had a deep connection with its advances in art, in literature and in those humanistic ideas that add wisdom to knowledge and form the basis of what we call civilization.

DECLINE OF GREECE TO GALILEO

For the purposes of this survey, the next thousand years were relatively uneventful. The decline of Greece is now frequently cited in heated academic debates as the inevitable failure of a department overly dominated by abstract theorists. There followed the counterexample of Rome, a nation of engineers and technicians who ignored everything that wasn't obviously useful. If this outrageous generalization wasn't enough to put Greek science on ice, Roman hegemony was replaced by the early Christian period where total acceptance of doctrine, obsession with love and charity tempered by theological expediency, stifled European science until the Renaissance. Greek philosophy was preserved by the Arabs. There took place some assimilation with Indian mathematics and astronomy. The Arabs also added considerably to the store of scientific knowledge until about the fourteenth century. Invention, arts and crafts continued, during the dark ages, to provide tools for the next scientific surge. The revival of scientific spirit in the sixteenth century brought a new element reminiscent of Archimedes: the manipulation of nature in order to illustrate and test rational models. Clearly, if theory were to be tested and inspired by experiment, the questions would have to be much more sharply drawn than was done by the global philosophy of the Greeks. This new mood is generally associated with Galileo and represented the birth of experimental science in close support of theoretical models.

The revival of science in Europe is represented by Copernicus, Brahe, Kepler and Galileo in a fruitful century and a half from about 1550 to 1700. Copernicus suits our examples well since his heliocentric theory accounts for the same data as does Ptolemy's model of the complex motions of the planets viewed from a central, stationary earth. If two apparently wildly different physical pictures account for the same data, how can one choose? Echoing the ancients, Copernicus wrote that "nature prefers simplicity in all of its manifestations," and casting *our* earth as just another planet (Copernicus didn't go that far, Kepler did) in solar orbits removed a fatal block to the progress of cosmology.

Here is a good time for reflection. Let the reader stand on a high mountain or on the bridge of an ocean vessel. Look at the night sky and you can't fail but to be impressed with the commonsense pre-Copernican idea of an infinite half-sphere as the vault of the heavens and a hemispherical, stationary earth floating below. However, as the human senses are sharpened by instruments and as the observations are set into mathematics, a deeper simplicity emerges. These are first seen by geniuses like Copernicus, Einstein and Bohr. They require from us mortals either much hard work or trust that uncommon sense must triumph. Now, still looking at the night sky but, holding tight to your perch, imagine a spinning earth, whirling through space around a sun you can't see. If you find *this* uncontested truth difficult to incorporate into your intuition, just wait till we get to relativity and the quantum theory!

The revolution produced by Copernicus produced no new

observational data. Ptolemy was the basic reference. The experimental side was taken up by Tycho Brahe (about 1570), who made measurements of great precision and provided Johannes Kepler with the facts that enabled him to quantify the observed motions of the solar system in terms of three simple, mathematical laws. The suggestion that these movements were the results of a force radiating from the sun was there, imprecisely stated but waiting for Newton's insight to be included in the monumental law of gravitation, which was to come.

Galileo was the first of the truly modern scientists. His use of the telescope yielded proof of the Copernican model and led to the conclusion that celestial and terrestrial phenomena have the same nature — an essential unification idea. Thus, the moon also had mountains, Jupiter also had a moon, four in fact, the sun rotated just as Copernicus said the earth must, and the earth must in order to comprehend the daily cycle of the firmament.

Galileo's discoveries in astronomy would have guaranteed him immortality. But they were as nothing compared to his creation of the modern style of scientific research. This is seen in many ways: (1) his taste in the phenomena to be illuminated; (2) his specialization in terrestrial mechanics, in falling bodies, inclined planes and pendula; (3) the style of precise measurement and mathematical description, (4) the crystalline clarity of exposition (he is probably *the* outstanding popularizer of science!). There was a sympathetic recreation of the best of Greek natural philosophers. Galileo's great books can be read with profit today by college freshmen majoring in humanities or business who are curious about how science works.

CALENDAR LEAVES, 1600–1900

One can tell the story of these years in terms of the triumph of syntheses; that is, the encompassing of a broad range of diverse phenomena by a single law of nature. Thus Newton, in his universal law of gravitation, understood that the same force governs the free flight of projectiles, including apples, on the surface of the earth, as impels the moon to orbit the planet.

By using Galileo's measurement of the motion of falling

bodies — 16.1 feet per second — Newton calculated the period of the moon's orbit as 28 days. The publication in 1687 of Newton's *Principia*, containing the universal law of gravitation, may well be the single most important event in the history of science (so far!). Newton's theory clarified the orbits of planets and comets, the bulge of the earth at the equator, and ocean tides. Later physicists would deduce that Newton's formula was valid throughout the galaxy and even between galaxies. Today, at Mission Control, the computers at NASA use this same equation to plot the trajectory of man-made satellites pulled in a variety of directions by the gravity of the sun, the moon and nearby planets. Experiment was important, and these scientists made many inventions as spin-offs of their research, such as telescopes, airpumps, thermometers, and optical spectrometers. Newtonian mechanics was spectacularly successful in explaining astronomical mechanisms and optimism was high that the human mind was competent to predict the future if only one could know the positions, velocities and masses of the objects of the world at any given time.

The nineteenth century continued the pace of science and of technology, and here we begin to see a closer coupling of these disciplines. Faraday was a preeminent scientist but his experiments in electricity and magnetism led directly to the dynamo and, more generally, to the electrical industry. Newtonian mechanics was carried to a deeper level of mathematical elegance by Lagrange and Hamilton. The latter perfected a comprehensive theorem of dynamics, called the principle of least action, which has survived into the modern period of relativity and quantum theory. It is much beloved by historians of science since it can be traced to Aristotle.

How fared the atoms of Democritus? The basic ideas were transmitted via Epicurus to the Roman philosopher Lucretius and through that connection to the Renaissance scientists, including Galileo. Newton used these hard, massy atoms in his work on chemistry and optics.

In about 1800, a Jesuit priest, Roger Boscovitch, wrote presciently about the primary elements of matter as mathematically pointlike — having no extension in space but acting upon one another by forces that he describes in detail: strong repulsion when the particles are close, attraction when they are more distant.

Experimental techniques for studying matter were improved by chemists, such as Lavoisier, who began to clarify the idea of chemical elements. Dalton in about 1800 summarized and advanced the idea. An atom is the simplest structure that contains all the properties of an element, and Dalton recognized twenty elements. This number would grow to ninety by the end of the century.

A suggestive step in the process of understanding the atom was made by the Russian chemist Mendeleyev, who discovered chemical periodicity and invented the charts that adorn all chemistry classrooms today. Prout, aware of the simple relation of atomic weights, suggested that all elements were made of hydrogen, recalling Thales and his contentious students. That a repetitive pattern in properties of elements would suggest a complex and similarly repetitive structure of the atoms corresponding to these elements came much later. The model of atoms, hard, indivisible, but subject to Newton's laws, led to new ideas about heat and energy. The experiments and ideas of Helmholtz, Joule, Thompson, Carnot, Rumford, Bernoulli and many others gave rise to the law of conservation of energy, to the concept of heat as the kinetic energy of restless atoms, to statistical mechanics — when there are too many atoms to keep track of, we treat them actuarially. Here again, with the application of the idea that all matter is made of atoms, a second great synthesis took place. The jiggling and chattering of atoms bombarding vessel walls explained "pressure." The increase in temperature of a gas is simply an increase in average velocity of the atoms (faster jiggling!). Heated liquids evaporate because their atoms move fast enough to escape. An enormous collection of observations on the properties of solids, liquids and gases became understood by this kinetic theory, much of it developed by a young English physicist, James Clerk Maxwell.

Conservation of energy was a primary unifying concept giving rise to the idea of a universal principle valid in all fields: in chemistry laboratories, in mechanical engines, in electricity, in biological systems and in the cosmos — in all processes — the total energy is constant. Other physical invariants were soon identified: angular momentum and linear momentum. Mathematical physics would later turn these overarching conservation laws into symmetry principles relating to the structure of space and time.

Another important body of experimental data came from the analysis of light emitted from glowing chemicals. The wave nature of light had been established by Young and Fresnel. Color was related to wavelength. The characteristic spectral lines found in the analysis of this light were not understood, but Wollaston, Fraunhofer, Balmer and others advanced the precision with which one could measure and fingerprint chemical elements. The observation of similar spectral emissions from the sun was an important advance in the cosmology of the nineteenth century. We now knew that the sun was composed largely of hydrogen and helium. This completed the work of Galileo and Newton in demonstrating that the things of the earth are the same as those in the sun, in the planets and in the distant stars. The study of moon rocks is an elaborate continuation of the sequence set out here.

But the most far-reaching advance of the nineteenth century was the synthesis, made in 1860 by James Maxwell, of the empirical laws of electricity and magnetism gathered over the previous 150 years. These were considered to be unrelated phenomena; magnetism appeared with lodestones and currents flowed in wires. Electricity had to do with rubbing amber rods to produce "charged" objects. These phenomena were known since Thales but now, mathematical descriptions of experimental results were available. Some of the experiments that provided the data were obtained by Coulomb, Cavendish, Orsted, Ampère and Faraday.

Out of this study came a clarification of the concept of forces acting between two separated objects. Faraday was the first to suggest that this force arises because of a field, created by the existence of electric charge. Faraday went on to a major discovery. Electric fields were not only created by charges but also by changing magnetic fields. The two heretofore different forces were connected. But the seminal breakthrough was still to be made.

Maxwell discovered that the data, formulated and reconciled by his mathematical equations, produced a permanent marriage of the electric and magnetic fields. A more remarkable result followed: not only did changing magnetic fields produce electric fields but changing electric fields produced magnetic fields. This implied the existence of self-sustaining and moving electromagnetic waves. These waves were soon identified with *light*. The entire spectrum of electromagnetic waves now stretches from the ultra-low frequencies and hence long wavelengths (kilometers) through the infrared, the visible region to the ultra-short waves (10^{-14}cm) radiated by excited nuclei and beyond.

We now had two force fields capable of acting through great distances: gravitation and the electromagnetic field. The unification by Maxwell of three historically diverse phenomena — electricity, magnetism and optics — led to a deeper understanding of the phenomena and was an inspiring lesson for what would come later.

AT LAST: TWENTIETH CENTURY

Following the instructions from Miletus, the twentieth century began with a robust discovery set. In 1897, J. J. Thomson identified the electron as one of the constituents of the atoms of all elements. He measured the mass (1/2000 the mass of the hydrogen atom) and the negative electric charge. Now the periodic table, the spectral lines and the existence of electrons together dealt the death blow to the idea of an indivisible, structureless atom. Curiously, the discovery of the parts that make up the atom also demolished the considerable resistance to the idea of atoms that still survived in 1900. In 1911, Rutherford sorted out the structure of atoms by discovering the atomic nucleus. This small central cluster carries most of the mass of the atom and enough positive charge to attract and neutralize the orbiting electrons. Thus, in Cambridge and Manchester, the atom was exposed, finally establishing its complex and distinctly nonelementary structure. Far from being an impenetrable sphere, the atom was seen as a greatly extended object, mostly empty space, containing a collection of internal constituents and internal properties. However, the atom also proved to be a window on a new and totally unanticipated level of truth. To cope with this new domain, a profoundly new concept of the world would be necessary. This is the quantum theory, which evolved from the first clues before 1900 and reached its modern form by 1930. Before we were through, the quantum theory would give us a total comprehension of atomic structure and atomic processes, reshape chemistry to a more profound science, explain the behavior of different forms of matter, such as metals, semiconductors, liquids, magnetism, superconductors and plasmas. Its technological consequences are vast and still growing. In philosophy, the impact was sharp, with new insights into the question of determinism and free will.

THE QUANTUM WORLD

In science fiction, earthlings descend cautiously from their spaceship, fully expecting a strange world, even one in which the laws of physics may be totally different. This is what happened when experimentation began to reveal data from within the atom. A dramatic example is the study of light emitted from heated atoms. One observes sharp spectral lines as if the atom were vibrating, ringing with sets of pure notes. This was evidence of some kind of resonant system inside the atom. How was this light generated? Rutherford's picture of electron orbits about a very small, massive nucleus ran into immediate problems with the physics of Newton and Maxwell. According to Newtonian mechanics, the electron must execute planetarylike orbits to have a stable system. According to Maxwell, an orbiting electron should radiate electromagnetic energy and soon spiral into the nucleus. Many efforts were made to reconcile these ideas. A new physics was operating here. Other experiments, examining the effects of scattering of light from electrons, verified a conclusion of Einstein that electromagnetic waves actually came in small packets of energy called photons. Thus the radiation from excited atoms was in the form of photons. However, the wave theory of light had been well established by the nineteenth-century stalwarts Young and Fresnel in definitive experiments illustrating the ability of light waves to interfere, crests of one stream of light waves falling on troughs of another stream to produce typical patterns of light and dark. How can this be reconciled with photons?

The resolution of these and other similar crises, pouring out of the new studies of atoms, came in a series of jubilant, daring, controversial contributions made largely in the period 1920–1930. Names like Bohr, Einstein, Heisenberg, Schrodinger, Born and Dirac were prominent.

The starting point is the recognition of an intrinsic and fundamental *inability* simultaneously to measure, with arbitrary precision, the position and the velocity of atomic particles. The Heisenberg uncertainty relations specify details of which quantities go together and what are the precision limits. These uncertainties are in turn derived from the inherent and unavoidable interference between the atomic scale object being measured and the measuring instrument, perforce *at least* atomic scale. For example, a precise measurement of position inevitably produces an uncontrolled disturbance of the particle's velocity.

So we start with *intrinsic* uncertainty as to what is knowable. It follows from this that the future evolution of an atomic system cannot be precisely traced. One can only compute a probability of finding the object in a certain state. The equations carrying the probability have only discrete solutions so that allowable states are discrete or quantized; only certain energies are physically allowed, only certain orbits permitted. The orbits are not now trajectories but cloudlike indicators of more probable locations. Nature does not contain any other possibilities. The discrete states are characterized by simple integers or quantum numbers; for example, the allowed energy levels of atomic hydrogen are given by the inverse squares of a set of numbers: 1, 2, 3, . . . times a basic unit of energy. And it worked! A vast amount of previously unassimilated data was now accounted for. And in the quantum theory, number is pervasive. The theory made hosts of new predictions that were quickly verified, often with great precision. The paradoxes of waves and particles were artifacts of carrying our humanoid experiences into the microscopic domain. Check them at the door, we learned. In the atom, we would find this mélange of uncertainty and precision.

To the scientific community of 1925, these ideas came very

hard. Einstein, in spite of his early contributions, never became reconciled to the ultimate interpretation. To a modern graduate student, after several years of hard work, quantum notions became part of a new intuition that must be acquired. Resistance to the new idea crumbled with success after success in accommodating to previously inexplicable data. It was helpful to have Bohr's brilliant demonstration that the quantum equations go smoothly over to the classical Newtonian ones as we increase the masses and numbers of atoms. The sharp determinism of classical physics is simply a result of probabilities becoming so high in the macroscopic domain as to amount to virtual certainties. Maxwell's electric and magnetic wave fields are similarly instructions as to where the photons must go. Whereas only the probability of the result of an observation may be predicted, the choices as to the final outcome can be calculated with great precision.

Perhaps a crude analogy will help. Consider a number of boxes each located with exquisite precision. Over each box there is a spigot. The spigots emerge from an overhanging spherical bowl the innards of which are hidden from us. We throw a marble into the bowl and cannot predict into which box it will land. But we can predict very precisely the possible final locations. Don't despair! This abstract piece of uncommon sense will evolve and be worth billions!

Dirac was the first successfully to combine relativity with the quantum theory. He applied the resulting theory to the electron moving in an electromagnetic field. Out of the resulting equations came the prediction that the electron was a spinning magnet. Tests soon confirmed this. Another result of Dirac's equations, much to the creator's astonishment, was the prediction of the existence of another electron — rather a twin particle of opposite charge and opposite sense of magnetism. The discovery of the positron in 1932 was a triumph for the relativistic quantum theory and an opening to the world of antimatter. In these days (as we will see later) of hot and cold running antiprotons, it is difficult to imagine the resistance and then the awe to which the concept of antimatter was subject. In the 1940s, profound but experimentally subtle refinements were made by Tomonaga, Feynman and Schwinger.

The result, quantum electrodynamics (QED), was capable of making astonishingly precise predictions. The magnetism of the electron can be calculated from the theory precisely and compared to experiment. It is expressed in terms of a parameter g.

g = 2.001 159 652 460 QED Theory

g = 2.001 159 652 200 Measurement

Note that a difference shows up only in the tenth decimal place!

Quantum electrodynamics seemed to be a powerful theory, valid in the macroscopic world (where it reduced to Maxwell's equations) and, in the microworld, compiling success upon success so that it was to become a stencil for a class of theories (quantum field theories) destined to impose order on a vast new complexity, undreamed of in the 1940s.

In summary, the quantum theory explained, with mathematical precision, aspects of the world that had, in the previous century, been considered outside the realm of science. Its applications to chemistry, biology, nuclear and materials science would contribute a significant fraction of the 1980s GNP of several trillion dollars.

RELATIVITY

Concurrently with this revolution in our conception of matter (but much more abruptly), there took place a profound change in our comprehension of space and time. This was the special theory of relativity, published by Einstein in 1905.

The special theory of relativity was also a response to the application of invention to fundamental observations. The observations created a crisis in the theory of propagation of light, understood now to be an electromagnetic disturbance. The ability to measure the velocity of light and to compare two round trips over distances of a few hundred meters, seeking differences of the order of one part in 100 million, were just not available to Maxwell, although he tried. Later, Michelson succeeded in this type of measurement and in showing that the velocity of light is the same whether the source of light is moving or not. There is a great deal of fascinating uncertainty about just how much this experiment stimulated the special theory. Nevertheless, Einstein's relativity analyzed and gave new meaning to physicists' measurements of space and time with rulers and clocks. Einstein's starting point (ex post facto or not) was a philosophical conviction that surely the laws of physics should not depend on the state of uniform motion or the position (or the race, creed and country of origin!) of the scientist-observer who is doing the experiments. Stated another way, the key to special relativity is that all observers, regardless of their special locations or states of motion, should discover the same laws of nature. This plausible premise led to a set of surprising consequences in conflict with common sense: for example, moving clocks run slow and rulers shrink when in motion. Relativity applied to dynamics gave the famous equivalence of mass and energy, $E=mc^2$. This fateful relation follows directly from the equivalence of all observers. It explained the binding energies of atoms and nuclei and provided the world with nuclear weapons. The special theory also permits the construction of cyclotrons and synchrotrons (which, in turn, advance science, cure cancers, make isotopes, etc.), and it is only when the quantum theory embraces relativity that the calculations of quantum states of rapidly moving particles are fruitful. The special theory of relativity introduced modifications in the Newtonian laws of motion. These modifications have observable effects only when velocities are a significant fraction of the velocity of light. As humans, we have no experience and therefore no intuition about such extraordinary velocities. We see again how sensitive instruments (Michelson's interferometer) lead to insights that apparently violate our commonsense experience but that contain a deeper elegance and truth in the description of nature.

GENERAL RELATIVITY

By 1906 Einstein had reason to be pleased. The previous year he had not only solved three outstanding problems of his day, only one of which was special relativity, but also was deep at work trying to understand the influence of his breakthrough as applied to Newton's law of gravitation. He had solved the problems in electromagnetism, reformulating the equations so that they do not depend on the position or velocity of the person making measurements. This seemingly created problems for gravitation where, according to Newton, the gravitational force depends on the masses and separations that do depend on the observer's state. The answer, published in 1915, was a new theory of gravitation: the general theory of relativity. Here again the philosophical key was the role of the observer.

In the new view, the presence of mass produced a distortion or curvature of the space near the mass. Objects chose the shortest path in the curved space. Thus gravity becomes a feature of the (non-Euclidean) geometry of the world.

The experimental verification of the general theory was not long in coming. The major influence of general relativity on our developing worldview is (probably) still in the future. This is because as yet there is no quantum theory of relativity.

General relativity did for gravity what the special theory did for electromagnetism. The most dramatic impact was in its application to cosmology; with relativity the oldest of sciences was now ripe for dramatic advances.

Einstein, in the period after 1915, set himself two major goals — the application of relativity to cosmology and the unification of gravitation and electromagnetism. Both efforts were unsuccessful, in different ways. Whereas general relativity predicted the big bang, black holes and the expanding universe, it was others who had the joy of reading these results out of the master's blueprint.

As early as 1901 the twenty-two-year-old Einstein, working on the problem of molecular forces, was guided by the powerful analogy with gravitation and, in 1901, he wrote about the "inner relationship of molecular forces with Newtonian (gravitational) forces at a distance." The molecular forces were, of course, complicated manifestations of the electrical force.

In 1925, only two forces were known to be operative in nature: gravity and electromagnetism. Einstein spent the next thirty years of his life in a futile effort to find a unified basis for these two forces. The reasons for failure lay in two directions. In the 1930s two new forces were discovered to play important roles in the universe. Then too, forces had to be treated in accordance with the principles of the quantum theory, which Einstein did not accept. Nevertheless, Einstein had a contagious faith in the existence of a deep, overarching unified theory that would provide a complete world picture: "... the story goes on until we have reached a system of greatest conceivable unity and of greatest paucity of concepts." Thus the Milesian ideal is restated with blinding clarity by the greatest genius of the twentieth century.

Einstein's effort at unification is the bridge to current work on grand unification, and to set the scene for this we must review the events of 1950–1984. The name of the subject we have been tracing is now high-energy elementary particle physics. The objective? Please refer to Sages of Miletus.

PARTICLE PHYSICS: THE BEGINNING

For a brief period, in the 1920s we had what seemed like a simple and elegant system. The world was made of just two elementary particles: the proton and the electron. These were collected by the electromagnetic force to form the atoms. Atoms aggregated to form molecules of infinite variety and hence to provide the vast array of substances out of which macroscopic objects were formed. On a scale of the solar system and larger, the gravitational force kept order in the universe, shaping orbits of the planets and comets and reaching out to gather stars into galaxies and galaxies into clusters.

Too simple. There were many open questions with this model. What produced radioactivity (discovered in the 1890s)? Where did the sun get its seemingly inexhaustible supply of energy? What kept the protons together in the larger nuclei in counteraction to the strong repulsive force of like positive charges?

In any commentary on the quality of reigning theory, open questions exist. The issue is whether one can find solutions within the constraints of the theory or whether these problems grow to crises that require radical revisions. We have seen that there was no way, within the scope of the Newton–Maxwell classical theory, that one would understand a stable hydrogen atom as modeled from the experiments of Rutherford. The solution was a radical alteration of the worldview.

And so it came to pass with the brief euphoria of the twenties. New particles were discovered, the neutron and the positron in the 1930s. A neutrino was proposed in order to make sense out of radioactivity experiments. The mechanism for radioactive decay of certain substances was interpreted as generated by a new force of nature. This was named the *weak force* since radioactive decay is a slow process. The puzzle of the stability of nuclei, tiny clusters of neutrons and protons, was settled by experiments showing definite departure from the inverse square law that characterized the electrical force. Thus the *strong force* was recognized, perhaps most clearly by Heisenberg. This would click into place when the nuclear particles came very close together and overwhelmed the electrical repulsion. However, not all particles were sensitive to all forces.

Uncertainly armed with the quantum theory and a score of ingenious experimental devices, physicists in the 1930s began the study of the nucleus. Naturally found radioactive substances were soon replaced by accelerators as instruments to probe the nucleus. The machines evolved from the electrostatic accelerators to the cyclotron of E. O. Lawrence, invented in 1930. Out of this research would come nuclear fission, a quantitative picture of nuclear forces and, eventually, a study of the shell structure of nuclei. As the energy of the machines increased in order to make more incisive observations, new objects, astonishing in their variety, appeared in the collisions.

Soon, this research merged with the work of a quite independent community whose interests were in the cosmic rays. This natural radiation was discovered in the 1920s and eventually identified as a particle bombardment, incident upon the earth from outer space. Particles of enormous energy, colliding with the nuclei of the atmosphere, produced a rich harvest of collisional debris. In the period 1930–1960, several new particles were identified by this technique. One was a very penetrating component eventually traced to a particle that weighed about 200 times the electron and was named the muon. This new particle was soon found by cosmic-ray scientists to be impervious to the strong force. Thus it was similar to the electron, which, in the innermost orbits of heavy nuclei, spends a major fraction of its time within the nucleus without noticeable effect. The aforementioned neutrino was also in this class. These were collectively named leptons, objects that ignored the strong force. Another important cosmic-ray discovery was a particle called the pion, generated in nuclear collisions and intensely sensitive to the strong force. This was an important breakthrough because such a particle had been predicted by the Japanese theorist Hideki Yukawa. Yukawa pioneered a class of theories of the new forces. Somewhat guided by Enrico Fermi's theory of the weak force, Yukawa proposed the pion as a carrier of the strong force. The 1980s version of these ideas will be treated below.

The discovery of new and presumably elementary particles, a trickle in the thirties and forties, was to become a torrent in the 1950s because of an application of wartime technology to the construction of particle accelerators. Equally important was a new style that came out of the war. Physicists had learned to manage large technological enterprises. After developing radar and the atomic bomb, constructing particle accelerators was not too difficult. Accelerators were needed to capitalize on an idea used so fruitfully by Rutherford. This was the technique of "scattering," a tool for probing the structure of nuclei. Philosophically, the accelerator was akin to a powerful microscope where accelerated projectiles replaced light beams. The scattering process consisted of directing a stream of rapidly moving particles toward the object to be studied and examining the results of the occasional collisions that took place. Generalizing from such experiments, one could glean information on sizes and shapes of the colliding particles and on the nature of the forces between them. One surprising result was that, from time to time, a new particle was created in the collision. This new object was thought to be summoned by the relevant force out of the energy carried in by the accelerated particle and in accordance with the relativity prescription, $E=mc^2$. Bring in enough energy (more than E), and it is possible to manufacture a particle of mass m. And new particles were made!

As the power of accelerators increased, a veritable zoo of new particles were identified and cataloged in accordance with measured properties. What are the relevant properties that define particles? We have mentioned electric charge, mass, radius. Particles are capable of spinning (like a top), and, since they contain charge, this gives rise to intrinsic magnetism. Particle properties reflect the quantum rules so that electric charges come in simple integers (quantum integers) times a basic unit of charge.

For example:

charge of neutrinos and neutrons: 0
charge of electrons antiprotons: -1
charge of positrons, protons: $+1$
charge of alpha particles: $+2$
charge of omega particles: -2

(etc.) times a unit of electric charge.

Other particles, to be discussed below, have numbers that are multiplied by one-third of a unit of charge. No other quantities of charge seem to be allowed by nature. As mentioned earlier, particles are capable of spinning, and this can also be measured. Here again only certain spins are allowed. These again come in integers that multiply a basic unit of spin. We have no such rules for mass, and this will come up later. Other quantum numbers further define the nature of particles: are they subject to the strong force? (If so, they are hadrons; if not, they are leptons.) Numbers are assigned whenever a regularity is observed, and a rule is needed to describe this. For example, leptons observed in countless reactions seem to preserve their "lepton-ness." This can be characterized quite neatly if we assign a lepton number, L, to each lepton, $+1$ say, and assign a -1 to the antilepton. Then we can pontificate a law of nature: "In all reactions, L does not change." So, lepton number is part of the description of particles.

Another important quantum number is carried by one class of hadrons. It is called the baryon number. The proton and neutron have $B = +1$. The antiparticles have $B = -1$. A lepton or a pion has $B = 0$ (they are not baryons). "In all reactions, B is also conserved" summarized the results of many observations.

In the period 1950–1970 some hundreds of new particles were identified in experiments with accelerators of ever-increasing energy. Here we should introduce the quantitative measure of energy (equivalent to mass). The unit is the electron volt (eV). The photons emerging from glowing atoms have a few eV of energy. X-ray machines and TV tubes can produce electrons of energy of the order of thousands of eV (KeV). The first cyclotron, invented in 1930 by Lawrence, accelerated protons to about 1 million eV (1 MeV). The mass of the electron is 0.5 MeV; that of the proton is nearly 1,000 MeV. The energy of the 1984 superconducting accelerator at the Fermi National Laboratory near Chicago is 1,000 billion eV or 1,000 GeV (Giga-electron-Volts).

Powerful accelerators staffed by about 5,000 experimental scientists in laboratories throughout the world, working over the past thirty years, have vastly increased our empirical knowledge of the subnuclear world. It is not possible in this survey to do justice to the individual feats of ingenuity that sparked this experimental research. A majority of the Nobel prizes in physics were collected by this group in this period, which is relevant since the practitioners constituted only a few percent of the physicists. An array of theoretical physicists have been equally busy suggesting, confusing and interpreting the experiments. To this activity, we must add the work of experimental and theoretical cosmology, since our instructions from Miletus had to do with the universe.

What has come of this? In the first half of the 1980s we can rather neatly summarize all of the data alluded to above (as well as the data obtained in less expensive laboratories going back to Galileo in Pisa in 1610). The synthesis has a name: the standard model. It summarizes what we have learned in terms of quantum field theories as they developed in 1930-40.

THE STANDARD MODEL

We referred several times to the quest for simplicity as an intuitively felt constraint on a complete worldview. Convictions deepened as successes came. The ninety or so elements of the 1890s were replaced by electrons, protons and neutrons in the 1920s. The four forces of 1800 — electricity, magnetism, molecular (chemical) forces and gravity — were reduced to two by Maxwell. And the hundred or so "elementary" hadron-type particles of the 1960s were reduced to three by the 1964 proposal that they are all made of different combinations of new objects we call quarks. The 1964 synthesis of Murray Gell-Mann had the quarks subject to the strong force and clustering to make protons, neutrons, pions. . . . For example, a proton is composed of two "up" quarks and one "down" quark. A pion is composed of an up and an anti-down quark. Three-quark structures are called baryons. Two-quark structures are called mesons. Nature does not seem to make other combinations. Leptons are not made of quarks. They are also elementary.

The quark idea evolved but also proliferated in order to account for new data. The 1980s standard model has six quarks and six leptons. These objects, in the standard model, are considered *atoms* in the Greek sense: primordial, unchanging, indivisible, containing no internal structure and hence no radius. They have a rich set of quantum properties. Combining these objects with the relevant forces (quantum field theories) gives rise to all of the objects and all of the processes we recognize in the physical world. How do we describe the forces? We have already seen that forces are described by fields, an idea that had its fuzzy beginnings in Greece and its first modern expression in Faraday's description of the electrical field. Field theories today are subject to the rigid constraints of the quantum theory and of relativity. They are known as quantum field theories, and imaginative theorists have invented a large variety of these in attempts to describe our forces.

Which is correct? Quantum field theories have had a checkered history, because they come with a disease. If one is not careful, one can calculate the results of some measurement (that's what theories are for) and have the result come out not only wrong, but also infinite! The disease of infinities has been with theoretical physics for a long time and large rewards are available for remedies. Thus, in the 1940s, three Nobel awards went to the doctors (Schwinger, Feynman and Tomonaga) who found a cure for infinities in the quantum field theory describing the electromagnetic force. Since then, the notion has arisen that nature must be beholden to a proper theory that doesn't have to be "cured" but is intrinsically free of the disease of infinities. There is also a very strong intuitive constraint that the correct theory will be a theory that not only gives an account of the data, but also offers the prospect of unifying the forces. One such class of quantum field theories is known and is by far the front runner in the "ultimate theory" sweepstakes. Such theories are known as "gauge" theories. They possess a basic simplicity or, a word much beloved by physicists, a "symmetry" known as gauge symmetry. More about this later; the word gauge is merely of historical significance. The idea is that gauge symmetry imposes a new constraint on the equations describing the forces. Symmetry seems to have the property of being resistant to the infinity problem. The beautiful mathematics loses in translation to pedestrian prose, but the important implication is that, in this description of forces, new particles appear. They are called gauge bosons, and their mission is to transmit the forces between interacting objects. They are the quanta of the force fields, and gauge symmetry makes crisp predictions as to their properties. In the 1980s three of the four forces (gravity is the holdout) can be well described as gauge theories.

The gauge theory of the electromagnetic force is quantum electrodynamics (QED). The corresponding gauge boson is none other than the familiar photon or quantum of light (electromagnetic field) energy. The photon was proposed by Einstein in 1905 and its reality verified experimentally in 1923. Its mass is zero and it has zero electrical charge. It has one unit of spin. The strength of the electromagnetic force between charged particles is measured by the amount of electric charge. The range of the force, i.e. how far it reaches, is, by the Heisenberg relations, determined by the mass of the force carriers, the photon. Since this has a mass of zero, the range of the electromagnetic force is infinite. The famous "inverse square law" is a detailed consequence of QED and tells us that the force between two small charged particles decreases as the square of the distance.

The strong force is now described by assigning to the quarks the property of "color," the analogue of electric charge. Electric charge is simply a measure of the electric force. Since this force has the simple property of attraction or repulsion, two varieties of this measure were needed. Whereas electric charge comes in two kinds (plus and minus, we call them), color charge comes in three types. This suggested the word color to someone, since there are three primary colors. Why three? This was simply the consequence of the experimental data. The gauge theory has therefore been named quantum chromodynamics (QCD). The gauge bosons are now eight in number and are called gluons. There is a gluon that connects red quarks to blue quarks and so on. If the reader adds them up and concludes there should be nine, he is correct. A delicate subtlety reduces this to eight. So quarks interact with one another by exchanging gluons. These ideas did not rise, full blown, out of thin air. There were so many data to be confronted and so many puzzles before the idea of color appeared to make the quark picture quantitatively satisfactory. Detailed calculations also showed that the original data, stimulating the idea of a strong force between protons and neutrons, are now thought of as residues of the strong exchange of gluons inside each proton and neutron. Gluons also have spin one and mass equal to zero. These properties have been verified by experiments in 1978.

The weak force is somewhat different. The quantity that corresponds to charge or color is not easy to describe, but it is a gauge theory and the gauge bosons are called W+, W– and Z° (Z-naught). These are very massive force carriers, and this contributes to the weakness and very short reach of the weak force. It wasn't until the invention of a new accelerator technology that experimental confirmation of the existence of these particles, one hundred times heavier than the proton, took place. This was in 1983–1984 and the place was Switzerland.

So far we have not discussed gravity. Although the general theory of relativity is indeed a kind of gauge theory, gravity is so feeble a force as to be totally irrelevant in the microscopic domain. Clearly it must be included in the overall worldview.

This then is the standard model, six quarks (each coming in three colors), six leptons. All of these have corresponding antiparticles, e.g., the anti-up quark has charge $-2/3$; the muon charge is -1, but the antimuon charge is $+1$. These primordial objects are subject to three forces via the exchange of twelve gauge bosons, i.e., the photon, eight gluons, two W's and a Z°. Sounds complicated? Let us remember that the model has had fantastic predictive power and does account for the data obtained in accelerator laboratories for the past 2,500 years. The model is internally consistent, not an easy thing to achieve in so vast a sweep of phenomena. However, the proliferation of fundamental objects, twelve quarks, twelve leptons, and forces with their twelve gauge carriers, has led to the notion that the standard model is not fully satisfying, in spite of its great success in describing the universe.

BEYOND THE STANDARD MODEL

How quickly we are rushing through this period! We pray that the reader will be as aware as we of the imminent end of this chapter and will vow to delve more deeply via a rich literature.

Physicists in the decade of the eighties are well aware of the edict of parsimony, even if they have neglected their Greek studies. The striving for simplicity comes with the métier. One branch of the attempt to simplify the model asks whether the quarks and leptons are indeed the end of the line. Could it be that they are themselves composite and made up of simpler (and fewer) objects? These hypothetical objects have been named prequarks.

The second front is the attempt to further unify the forces. This goes under the name of grand unification: i.e., the notion that the three forces we have briefly described are manifestations of a deeper, unified theory, just as electricity and magnetism were thought to be separate forces before Maxwell.

These are of course related tasks. For example, if quarks and leptons are the last word, we would still be impelled to look for unification. Alternatively, if prequarks exist, then there very likely will be a new force binding these into quarks and leptons, and this would have a profound influence on the unification quest. Recall Einstein's unification efforts, which failed without knowledge of the weak and strong forces.

To give even a remote sense of the unification drive we must discuss the notion of symmetry. Who hasn't seen the Greek temples not to be impressed with the influence of symmetry in art and architecture? Then, too, consider the beautiful patterns of shells or of crystals, now hard as diamonds, now fragile as a snowflake. It has been said that symmetry is one of the ways by which the human mind recognizes order in nature. In the domain of space and geometry, symmetry implies invariance — a key idea in expressing laws of physics.

Invariance: some quantity has the property of remaining constant in spite of things happening, processes taking place, e.g., collisions and so on. For example, the total energy in any isolated system is constant. The system could be a swarm of particles in random collisions, coalescing, disintegrating, forming crystals, evolving, tumbling, here boiling, there freezing. The law of nature that insists that in all of these dynamic processes the total energy is a constant implies an order. Once this is appreciated, the order can be used to make predictions.

The speed of light is constant, as is the total electric charge and so on. Consider a long row of Doric columns and think to shift the row by one column spacing. Nothing appears to have changed. The situation is invariant to the shift. Rotate a perfect sphere (pure symmetry!) by any angle about any axis, and no one would know it was done. It looks the same. Displace a laboratory by seventeen kilometers from point A to point B. The symmetry of space guarantees that all the laws of physics discovered in this laboratory will be unchanged. The search for the deep laws of nature has become a search for symmetry. In today's textbooks, the laws of conservation of energy, momentum and charge are all expressed in terms of symmetries not all confined to geometry and space. We have alluded to the gauge theories, i.e., quantum field theories that obey gauge symmetry. What was spectacular was that the hypothesis of gauge symmetry implied the existence of force. The first example was the imposition of a geometric symmetry on space-time by Einstein out of which he was able to deduce the gravitational force. We soon learned that the most elegant description of all the known forces was via their invariance with respect to this gauge property. Our archetypical, precise theory, quantum electrodynamics, was successful before it was realized that it was the perfect gauge theory, invariant to changes in the measure or gauge of the state.

The attempts to find a gauge invariant theory of the weak force led, in the 1970s, to a partially unified theory of the weak and electromagnetic forces. This breakthrough was a labor of many in the decade of the sixties. Progress was the result of the blending of sensitive experiment and resilient theory, each stimulating the other until the final synthesis. At first sight, the combined symmetry should yield four gauge force carriers, all of zero mass, two charged and two neutral. The experiments measured zero for the mass for the photon but large masses for the W's and the Z. How to account for this blatant asymmetry without destroying gauge invariance was the problem. There arose the notion of "symmetry breaking" or "hidden symmetry." Fundamentally it was a discovery that, whereas the symmetry of the mathematical equations describing the system exists, the solutions to the equations corresponding to low energies and low temperatures may be asymmetric.

A crude analogy would be to consider a spherical bulb filled with steam, i.e., water vapor at high temperature. This is pure symmetry subject to the laws of nature. When we lower the temperature enough, the laws of nature are such that the system develops ice cubes floating in water. The symmetry is apparently destroyed and a much more complex system appears merely as a result of the lowering of the temperature. The basic symmetry (or simplicity) is there but hidden. In this way, terms could be added to the theory that produced asymmetry (massive gauge bosons) in the lowest energy solution to the equations without destroying the basic symmetry. When this was done to the combined electroweak process, lo and behold! Three of the gauge bosons acquire mass and one doesn't. The theory was able to predict the masses of the W+, W– and Z° to an accuracy of about 1 percent and, in 1983, these predictions — the existence and correct masses — were brilliantly verified. The unification of electromagnetic and weak forces, with an added symmetry-breaking feature (this would come to permeate the efforts at a complete synthesis — it was known as the Higgs mechanism) was not really complete unification; the forces remained distinguishable. Nevertheless it greatly encouraged attempts to include the strong interactions into complete, so-called grand unified theories (GUTS). These, it was quickly pointed out, were not all that *grand*, since gravity is not included.

The restoration of symmetry in the electroweak theory is accomplished if the energy of the system is raised so high that the mass of the W and Z particles is negligible. Then, nature appears to be whole again, and the four gauge bosons of the unified theory are on the same footing.

In the GUTS theories, there is a defined energy at which overall unification is true. The basic idea is that the strength of the strong force decreases at high energy, whereas the strength of the electroweak force gradually increases. They meet at an energy of 10^{15} GeV, about 100 billion times higher than the highest energy accelerator now being constructed. At this energy the forces are truly unified: there is only one force; all particles are equivalent since masses are negligible; and complete symmetry (still neglecting gravity) reigns. Why is gravity so tough? Basically, theorists found that attempts to subject gravity to the quantum field mode produced terrible infinities.

PARTICLES AND COSMOLOGY

This is the appropriate time to bring in cosmology. Is there an objective reality to a regime consisting of particles with the enormous energy of 10^{15} GeV? There turns out to be a plausible one provided by the cosmological theory of the big bang. According to this consensus, the universe was created about 15 billion years ago in a cosmic explosion. In the earliest fractions of a second, the universe was very dense and very hot. At these temperatures, matter was divided into its most primitive constituents. As time proceeded, space expanded and the universe cooled. The primordial particles, in dense and violent collisions, eventually were able to coalesce. Thus the "atomism" of Greek science, through its evolution into quarks and leptons, is necessary for an understanding of cosmological processes. The connection between temperature and energy is almost a trivial one: 1 eV = 10,000 °K. Some figuring then shows that the grand unification energy is 10^{28} K. Thus, when the universe "cooled" to 10^{28} °K from its much hotter beginning, the strong, weak and electromagnetic forces were indistinguishable. Even the difference between quarks and leptons disappears. They are now generic spin one-half objects. As the universe continues to expand and cool, the forces become distinguishable, the symmetry "breaks" and a complex but perhaps more interesting world begins to form.

If quarks and leptons can mix intimately, quarks can change into leptons, and baryons (made of three quarks) can also change to leptons. Baryon number is then not conserved in the early universe. This is a basic prediction of detailed grand unified theories. It has implications for present-day experiments; if GUTS are right, protons should decay, however rarely, even in our cold universe. Early estimates of the proton's mean-life were of the order of 10^{31} years (so don't worry!). Can one measure this? The trick is to assemble enough protons and watch them for enough time. Watching thousands of tons of protons for several years has, by 1987, failed to detect the decay of the proton. The validity of GUTS is still an open question, in spite of great aesthetic appeal and some experimental successes. These are qualitative but nevertheless impressive.

The inclusion of gravity in unification of forces is a subject of very active work in the 1980s. There are formidable problems, but there is optimism. The name is supergravity, an attempt to bring the quantum theory to general relativity. The difficulty is that quantum gravity effects have yet a higher scale of energies (or temperatures) than GUTS. The energy is 10^{19} GeV!

Notice we have a mysterious but smooth hierarchy of forces. Consider again the rapidly cooling early universe. At temperatures so hot that all the particles have energies in excess of 10^{19} GeV, we have complete symmetry with quantum gravity included. The presently observable universe was then microscopic in size and had lived but 10^{-43} seconds when the act of cooling caused gravity to condense, leading to the GUTS epoch. By 10^{-35} seconds, the connected universe is about ten centimeters in diameter and the strong force separates from the electroweak. Leptons and quarks are distinguishable and clusters of quarks begin to make hadrons. Much later, 10^{-10} seconds, the temperature falls to an energy equivalent of 100 GeV (mass of W, Z) and the electroweak force comes apart, weak and electromagnetic forces separate, and our symmetry, like the flask of ice water, is only a memory. Fifteen billion or so years later we are in a universe of average energy one three-thousandth of an electron volt (3° K) and scratching our heads about where we go from here. The recognition that particle accelerators can organize collisions (one at a time, of course) that were characteristic of all the particles in the early universe created a symbiosis between these two subjects closer than any that had previously existed.

A complete understanding of cosmology must include quantum gravity, since it becomes as strong as the other forces at the very earliest instants of creation. In the theory of supergravity, the gauge particle, the graviton, which has two units of spin, is related to the lower spin quark-leptons (spin one-half) and gauge bosons (spin one) by a new symmetry, called supersymmetry, which indeed mixes particles of different spins.

Supersymmetry is a powerful and attractive procedure for writing down a theory without infinities, a theory that promises unification, too. Supersymmetry even manages to include the theory of gravity heretofore resistant to any form of quantum theory. Supersymmetry introduces a vast number of new particles since it installs a partner for each of our present particles. The twins of the spin one-half quarks and leptons have spins of one unit and are called squarks and selectrons, i.e., super partners. The partners of the bosons (spin one objects) are designated by adding "ino," i.e., photino, wino and zino as well as gluino. Since none of these objects has been as yet observed, they must be heavy. Since they are heavy, supersymmetry is "broken," i.e., the symmetry is hidden. How heavy? Heavy enough to have escaped detectors, which means that the lightest of these must be in the tens of GeV.

Theorists tend to clutch at straws for support of these very speculative ideas. One such straw comes, again, from recent astrophysical observations. Studies of the orbital motion of both astronomically sized bodies and of molecules around galaxies indicate the presence of matter that has no luminous properties, hence, dark matter. With supersymmetry, there are many candidates for materials to serve as dark matter. Another straw is even more exotic. It sticks down from the very misty outer limits of human speculative endeavor, the so-called theory of superstrings. The origins of this attempt at an overarching theory go back to the early 1970s, but its almost universal appeal to theoretical physicists (mostly those under thirty!) developed relatively recently when it was shown that this could be the best possible candidate for a unified theory of all forces with no infinities.

Superstrings theory assumes that primordial particles are not *points* in space but are short strings. Such theories, to be consistent, require more than three space dimensions (nine is the current favorite!), but it is taken that a symmetry-breaking mechanism curls up the extra spatial dimensions into unobservable domains. Part of the attraction for theorists is that it requires "new math," which makes a mathematician's eyebrows curl; also the physical domains where the theory readily applies is in the 10^{19} GeV domain — tough to test. The straw down to potentially observable energies is as yet tenuous but does tend to point toward the idea of supersymmetry.

THE FUTURE: SCIENCE

We have alluded to the open questions, and perhaps these should be reviewed. The standard model invokes a symmetry-breaking operation that gives the W's and Z° mass. This implies the existence of a new object: the Higgs particle, which has not yet been identified. Red-blooded experimenters are told: "Find the Higgs!" The GUTS need experimental guidance since the various attempts, like supersymmetry, make predictions that imply new, massive particles. We still do not understand the deep nature of mass, apparently a patternless array of parameters. The proliferation of quarks and leptons argues for a new layer of structure, prequarks, which are beyond the range of current inventory of accelerators. Cosmological support for the developing worldview is very encouraging but rests on the observations we can make today, billions of years after the fact. We need some closure of the gap between present accelerator energies and those unimaginably pure instants when the world was even more perfect than our Greek philosopher could imagine. For all these reasons, physicists are straining the state of technology in order to devise new and vastly more powerful machines to explore the higher energy domain.

Progress in the 1970s and 1980s was generated by the

construction of large particle accelerators of ever-increasing energy and cost effectiveness. In 1987, we can clearly see the experimental data that will begin to emerge from two new particle
accelerators in the United States and from two new machines that
will begin to operate in Europe by the end of the decade.

These machines will explore the domain at the limits of the
standard model. At Stanford (California) a machine, called SLC
for Stanford Linear Collider, which observes collisions of electrons against positrons at an energy near 100 GeV, will begin to
take data in the summer of 1987. It will be a "Z°" factory,
producing thousands of these (so recently!) exotic particles in
order to study all possible disintegration modes. The LEP machine at CERN, near Geneva, Switzerland, is a deep, underground
ring twenty-seven kilometers in circumference for containing two
counterrotating beams of e+e– up to somewhat more than 200
GeV for going beyond the Z° mass scale. The HERA (Hamburg)
machine is unique in colliding protons against electrons at a total
energy of about 350 GeV. In 1983, two technical breakthroughs
emerged to set the stage for what could be the "final assault" on
the summit of theoretical science. The first was the taming of the
technology of superconductivity with the successful operation of a
1,000-GeV superconducting accelerator. This took place in the
unlikely location of Batavia, Illinois, a former cornfield in the
American heartland turned premier particle physics laboratory.
Superconductors, based on exotic alloys, have the property of
conducting large electrical currents to create powerful magnetic
fields without the normal heating effects that make large electrical
power bills. This was technology with a vengeance and with more
than 700 microcomputers to control precisely the operations of
the world's most powerful accelerator, designed to illuminate the
details of quark properties. The second breakthrough was the
successful deployment of a tour de force of accelerator gymnastics
known as the "collider mode." This was carried out in the
European laboratory outside Geneva where twelve countries
pooled resources to give the Americans competition in a field that
was born and nurtured in Europe. The collider technique involved
the organization of head-on collisions of two counterrotating
streams of accelerated particles. This trick made possible observa-

tions of vastly higher sensitivity. The European effort led to the successful observations of the long-predicted W's and Z particles so crucial to the standard picture.

The "Tevatron," Fermilab's particle accelerator, is designed to organize collisions of protons and their negative twins, antiprotons, after January 1, 1987. The total energy is 1.8 TeV, a new world's record and well beyond the mass-scale in which nestles the standard model.

But the Tevatron success has had another effect. It has encouraged the possibility of another quantum step in energy, a step that has looked increasingly necessary with the theoretical developments of the early 1980s. By 1983, the urge to define the facility of the middle 1990s exploded in the U.S. community, and there arose what became known as the SSC proposal.

SSC stands for superconducting supercollider and has a design goal of colliding two counterrotating beams of protons at a total energy of 40 trillion electron volts.

The idea of combining the two breakthroughs became self-evident, and the hopes of addressing the open questions rest on a superconducting supercollider. European and Japanese versions of these dreams also exist. These machines may appear in the 1990s to address what could be the residual issues in a 2,500-year-old quest for a rational understanding of the physical universe.

THE FUTURE: SOCIETY

In the course of this essentially cultural activity of pure science, the practitioners have changed the world in all possible ways. In spite of all the scaffolding of the scientific methodology, the scientists have characteristically maintained the deep, human traits planted in our genetic and cultural heritage. We still impose intuition, elements of aesthetics, subjective judgments of beauty, in guiding the science. "Of course we don't believe in horseshoes but some people say they help anyway." Whereas, starting with Copernicus, we have gradually displaced Man from centrality in the cosmos, recent cosmological thought has brought him back via the anthropic argument — the notion that the laws of nature of this universe are what they are in order to generate conditions favorable to the evolution of intelligent observers. Be that as it may, the quest for a worldview will surely continue to impact upon society beyond culture.

The technology developed and developing in the frontier laboratories finds its way into commerce and works its changes. In the 1950s, the work at accelerator laboratories greatly influenced what is now the computer age. New materials and new technologies came out of the need for a sharper view of quarks and leptons. No one can predict the effects of today's exotica, like grand unification, on the ordinary citizen of the twenty-first century. Speculation is great fun, and one can weave tenuous fabrics: unlimited and cheap energy from conversions of baryon mass to useful energy; convenient energy storage via bottles of trapped antimatter; the mining of local black holes to produce enough of everything for everyone and so on. These are standard and extravagant speculations. The more solid achievements, which are derived from the process of acquiring the new knowledge, serve to pay the bills. In our history no one has yet succeeded in overestimating the impact of science on society. If the developing revolution in physics fails to have profound societal influences, it will be the first time in history. To make an objective judgment, past results have been rather positive although ominous possibilities for infinite mischief exist. Deeper changes may arise from the culture itself. After all, Newtonian science radically changed the nature of religion and caused the abolition of witchcraft. Spaceship Earth has dramatized a new environmental ethic; the computer age is affecting language and styles of thinking. Recall the extreme popularity, rivaling film and rock stars, of Albert Einstein in the 1920s and 1930s and the tens of millions of people who watch programs about science on television. A new worldview, given time to percolate into the consciousness of society, can have unimaginable consequences, even without hardware.

Chapter Two

Tomorrow's Thinking Computer

By Richard Flaste

Richard Flaste is the director of science news of The New York Times.

As news stories go, it did not get much attention in New York. The article, on April 14, 1982, was brief and buried deep in *The New York Times*. It was a dispatch from Tokyo: "The Japanese Ministry of International Trade and Industry has authorized the formation of a multicompany research institute to develop a Japanese supercomputer by the 1990s, the Kyodo News Service reported today. The ministry approved an application filed by six major Japanese computer makers and two electrical manufacturers to set up the research and development group, to be named the Institute for New Generation Computer Technology, Kyodo said. The research institute will receive a grant of $1.7 million from the Ministry of International Trade and Industry in the fiscal year 1982 for development of high-speed 'fifth-generation' computers within the present decade, Kyodo said. The group, headed by Takuma Yamamoto, president of Fujitsu Ltd., had invited United States companies to join, but no applications from American makers have been filed, Kyodo said. There was no immediate reaction to the report from Japanese Government or industry officials." That's all there was to it. The dispatch was a simple, formal acknowledgment of a movement that had been afoot for months in Japan and that had already begun to inspire concern, even dread, in other nations of the world where there were some who feared that information processing, the next great frontier, would be dominated by the Japanese and with it the world.

The Japanese actually intended to develop two kinds of advanced computer. One would be a "supercomputer" that would be able to perform numerical calculations far faster than the most powerful existing machines. The other would be the radically new "fifth-generation" or "artificial intelligence" computer that would have some reasoning power, be incredibly simple for the average person to operate and solve a whole range of problems that are not readily handled by numerical computers.

The twin programs became Topic A among American technical leaders in government, industry and the universities. Some Americans, especially, worried that their fragmented, inventive approach to computers, an approach that had kept them at the head of the field for decades, would fail them now, and their days of supremacy would end in the face of a meticulous Japanese assault. That fear was, in some measure, an obvious response to the stunning efficiency of the Japanese in postwar electronics. A country with 110 million people in an area the size of California had found an ally in the tiny electron, a subatomic particle that could be disciplined into one of humanity's most powerful and useful tools.

"The nation that dominates this information-processing field will possess the keys to world leadership in the twenty-first century," Robert E. Kahn, a top Defense Department computer expert, warned. And Joseph F. Traub, chairman of the computer science department at Columbia University, added to the concern when he remarked: "The significance of the [Japanese] project is that it could mean who dominates in computers in the 1990s and thereafter. And the nation that dominates in computers will be the dominant nation economically."

Other computer experts suggested that the United States

might soon succumb to a technological Pearl Harbor, possibly even reverting to an "agrarian society" that would provide food, but not computers, for its technological superiors in other nations. The pessimism and worry in the United States seemed well enough founded. A panel chaired by Peter D. Lax of New York University's Courant Institute said, "The .U.S. has been and continues to be the leader in supercomputer technology and in the use of supercomputers in science and engineering. . . . However, the panel finds that this position of leadership is seriously undermined by the lack of broad-scale exploration outside of a few national laboratories, of the scientific and engineering opportunities offered by supercomputing, and by a slowdown in the introduction of new generations of supercomputers. This threat becomes real in light of the major thrust in advanced supercomputer design that is being mounted by Japanese government and industry, and by vigorous governmental programs in the United Kingdom, West Germany, France and Japan to make supercomputers available and easily accessible to their research and technological communities . . ." The White House science office convened panels of experts to assess the situation, the House Science and Technology Committee held two sets of hearings on computers, the Defense Department launched a $600 million research program in artificial intelligence explicitly designed to counter the Japanese challenge, and the National Science Foundation and other government agencies set up programs to encourage greater use and development of advanced computers. Meanwhile, leading American electronics companies set up new research consortia to help keep them ahead, or at least abreast, of the Japanese. A similar burst of activity occurred in Europe; the British, the French and the West Germans all started or greatly accelerated government programs to develop advanced computers, and electronics companies boosted their own efforts. As the great computer race intensified, so too did the confusion. Observers were deluged with such a blizzard of claims and counterclaims, alarms and reassurances, ballyhoo and bunk that it was difficult to discern who was winning the race and what difference it really made. So many of those making pronouncements stood to gain or lose from whatever government actions might be taken that it was difficult to sort out whose assessments were objective and whose were motivated by political or economic self-interest.

But the Japanese phrase for this next push in electronics — fifth-generation computers — was catchy enough to have been coined on Madison Avenue. It conjured up something inevitable, a kind of manifest destiny, and yet its exact shape and function were unknown: thus, the allure of mystery. For some, the quest for the fifth generation had about it the same sort of romance that the ascent to the moon did decades earlier.

All the promotional razzmatazz very nearly masked a truth at the heart of the computer race. This quest was indeed one of nearly transcendent importance. Artificial intelligence in fact held the potential, if it could ever be developed, of changing the nature of life on earth in some profound way. In all likelihood, it would never replace the human mind; rather, many thought, it would emerge as a powerful adjunct to it. AI, as it was called, would provide mankind with a tool for dipping into a reservoir of stored knowledge to aid in faster and more efficient decision-making about nearly everything, from where to drill for oil to how to treat an ailment. Already "expert systems" were in use. Relatively "intelligent" computer systems were helping the telephone company troubleshoot problems, for instance, and some doctors were already turning to their computers for diagnostic reasoning.

The Japanese term, fifth generation, despite the aura that surrounded it, was mostly utilitarian. It referred to an often discussed hierarchy in computer evolution. First there was a generation of huge machines that relied on vacuum tubes; then there was one that used transistors; a third relied on the integrated circuits printed on silicon chips; and a fourth — a generation only now emerging — turns to very large-scale integrated circuits (VLSI), with chips so compact they must be designed by another computer. The achievement of artificial intelligence would be the crowning effort, the fifth generation.

The fifth generation would come dear. That was evident enough from the supercomputers that already existed in the earlier generations. Supercomputers were extremely expensive, selling for $20 million each and designed for large scientific problems involving numerous calculations. They performed in some case

American mathematician, came up with the basic scheme of the store-program computer, the essential blueprint of operation for computers had not changed. The computer's brain, the so-called central processing unit, executes arithmetic calculations one at a time before going on to the next calculations. Over the years, computers had become smaller and faster, but they still used the von Neumann format, or architecture, known as serial processing.

But the fifth-generation computer — whatever it eventually turned out to be — would have to be very different.

THE NEW GOAL: PARALLEL PROCESSING

In their quest for computers that are superfast and supersmart, scientists in Japan — and in every other industrial society that had joined the race — were renouncing the von Neumann principle. Instead of solving problems step by step, the new machines were intended to break apart computational puzzles and solve thousands and millions of their separate parts all at once. The revolutionary approach took on the unassuming name of parallel processing.

As parallel processing entered its infancy in the mid-decade, Dr. Kenneth G. Wilson, a Nobel laureate at Cornell University, said, "It's a crucial step; there are problems we'd like to solve right now that require 100,000 times more computing power than is currently available. The only way to conceive of getting it is with parallel processing."

"We're at a pivotal point in computer history," said Dr. Salvatore Stolfo, a computer scientist at Columbia University. "Fifty years from now people will look back on this transition with almost mystical respect."

Despite the enormous size of the step, the basis for it was rather simple in principle: two heads are better than one. Scientists want to combine two, four, sixteen, and possibly as many as a million individual computers to work simultaneously on a single problem. Over the decades, the stunning accomplishments of the von Neumann machines had come about not because of fundamental changes in the way they work, but because engineers had been able to cram increasingly powerful logic into ever smaller spaces. The upshot was increased speed.

But that upward march of increasing power and speed started to slow. Fundamental physical constraints were impeding the rate at which chips could shrink. Wires could be only so thin. Transistors on a chip could be made only so small. The constraints were forcing scientists to think about computers in entirely new ways, such as combining processors to work in parallel. And computer experts liked to point out that nature had provided important clues to parallelism as the right way to achieve machines that think. Human brains, after all, use parallel processing.

Dr. Tomaso Poggio, a physicist at the Massachusetts Institute of Technology's Artificial Intelligence Laboratory, observed that the brain's neurons operate about a million times more slowly than silicon. The reason that the brain nonetheless works so well, he said, is that the neurons in many cases work in tandem. "With the slow speed of the neurons," he said, "there's no other way to do the astonishing things it does."

A simple example of a parallel computation, Dr. Poggio said, comes from the quest for computer vision. A machine programmed to recognize the image of a horse needs to compare its stored memory to data coming from a camera. To get a match, it might need to increase the contrast of the incoming image to get a clearer outline of the real horse. A von Neumann computer, he said, would do this by performing a calculation on each data point of a certain value coming from the camera, a process involving hundreds and thousands of individual steps performed one at a time. A parallel computer, on the other hand, would do these calculations all at once. "The main advantage is speed," Dr. Poggio said. "It makes a huge difference if it takes a second or a day to solve a problem, especially for ones concerning motion or changing values."

Vendors of commercial computers had already started to dabble with parallel processing. Cray Research Inc. introduced a machine, the experimental multiprocessor called X-MP, which can operate with up to four processors in tandem, according to Robert Gaertner, a Cray vice president.

So, too, Denelcor Inc., in Aurora, Colorado, introduced a line of supercomputers that could combine up to sixteen individual processors. According to Darrel McGinnes, head of product management and planning at Denelcor, a more limited version of this sixteen-processor system had already been sold to the army's Ballistic Research Laboratory, the Los Alamos National Laboratory, the Argonne National Laboratory and a West German aerospace firm.

Beyond sixteen processors, the quest for parallelism enters the realm of pure research and a bewildering array of approaches. Von Neumann machines come in one form. But there are thousands of ways to try to build a parallel processor. At universities and in research institutions around the world, dozens of different approaches were under development. Scientists at the California Institute of Technology built a machine called the Cosmic Cube that hooks together sixty-four boards, each basically equivalent to an IBM personal computer. The overall machine has roughly one-tenth the power of the Cray 1 supercomputer at about a hundredth of the cost. At Columbia University, Dr. Stolfo was trying to build a machine, known as Dado, that will combine 1,023 separate processors. The machine, he said, is not as foreign as it sounds.

"There are lots of examples of parallelism in real life," he said. "Take an election. If it were done by a von Neumann machine, only one person could vote at a time. It would take months to cast all the ballots. But in the real world you distribute the processing power to individual states, counties, cities and districts that simultaneously record and count up the votes. That's exacly what parallel processing is doing for computer technology."

As the number of processors increases, they often become simpler, according to Dr. David Shaw, another scientist at Columbia who was pursuing research in parallel processing. His machine, known as Non-Von, ultimately may contain as many as a million processors all working in concert. "Crays at most use a few processors," he said. "The thing that distinguishes our work is that it relies on a much larger number of smaller, simpler and slower processors. We organize our machines in very different ways."

One challenge for parallel processing is flexibility. Because computers hooked in parallel often need to work on problems that are similar, their focus can become quite narrow. It is easy to build parallel processors that do one thing, such as solving equations for artificial vision, but it is difficult to make them flexible enough to take on a wide variety of tasks.

So too, "some problems do not lend themselves to parallel-ism," Robert Kahn of the Defense Department's Advanced Research Projects Agency (DARPA), a major source of funds for parallel processing, wrote recently. "Planning and reasoning appear to involve a sequential component, coupled with bursts of concurrent activity to explore options and alternatives."

Some researchers said they doubted that parallel processing by itself would bring about breakthroughs in artificial intelligence. "When you don't know what else to do you build hardware," said one computer scientist. But a scientist who saw parallel processing as the start of a new epoch is Dr. W. Daniel Hillis, a scientific founder of the Thinking Machines Corporation, an MIT spinoff. The Boston-based company is building a device called the Connection Machine, which is ultimately meant to have a million processors.

"Computer science is at a pivotal juncture," said Hillis, a young scientist who was just completing work for his Ph.D. "Nobody knows just what they're doing in the parallel area. But that's the nature of pure research. It's an incredibly exciting time to be in the field. Things that we only dreamed about are suddenly becoming possible."

Hillis may have put his finger on what was really happening. The excitement was palpable and real, but nearly everything else about this bewildering field was conjecture.

WHO'S REALLY AHEAD?

One of the biggest questions centered on national security. Was the national integrity of the United States threatened in some way? The question of where the United States, Japan and Europe really stood in the great race didn't start to come into focus until sometime in 1984, three years after the competition began with the kickoff of the Japanese national computer projects. A new wave of reassessments and informed judgments by technical leaders made it possible to see the computer race in better perspective. It was indisputably clear that the United States was indeed facing a serious challenge in computer fields where no challenge existed before.

But cool assessment suggested that the United States was still the world leader in computers, artificial intelligence and related scientific fields. The Japanese efforts looked somewhat less fearsome than originally thought. And virtually all of the grandly announced programs around the world were having start-up problems. There was a growing backlash to the alarms being sounded about the threat to the United States from abroad.

"There's a great deal of hype involved in all this, a lot of alarm and people jumping up and down," says Edward E. David, Jr., president of Exxon Research and Engineering Co., who was science adviser to President Nixon. "I think we've gotten overly concerned about what is effectively a research and training effort by the Japanese. I don't see that this will lead to worldwide dominance."

The official national programs were all roughly comparable in size, with most of them calling for expenditures in the half-billion-to billion-dollar range over the next five years. The fifth-generation budget, for example, was $430 million in the first five years, roughly comparable to the Defense Department's "strategic computing" budget of $600 million over five years.

By some yardsticks, those were significant sums. The American government would be pouring tens of millions of dollars a year into the nation's universities to increase the use of supercomputers in research, and the Defense Department would be channeling far more money into artificial intelligence research than ever before, vastly increasing the size of the field. But compared with the $3 billion or more that IBM spends on research and development in a single year, the new programs seemed moderate.

· The great computer race was far less costly and sweeping than either the arms race or the space race. Many experts likened it to the competition for supremacy in such other technical fields as electronic chips, communications, biotechnology or even airplanes and automobiles, with the exception that computers have more potential to affect a wide range of activities than do most other technologies.

Many experts said there had been enormous confusion generated in the minds of the public and of government officials by mixing up two different computer races — the one involving numerical supercomputers, the other involving fifth-generation machines — that were only loosely related. Each field was progressing at a different rate and had different prospects for success and vastly different consequences for military and economic power.

The fifth-generation programs offered the greatest potential for truly revolutionary breakthroughs that could change the shape of whole industries, social organizations and military forces, but the likelihood of such payoffs in the near future appeared remote. Faster numerical supercomputers, on the other hand, were almost certain to be developed in the decade, but they would be far less revolutionary in their impact.

Here's how the battle for supremacy in each area was shaping up.

SUPERCOMPUTERS

American companies have long dominated the design and production of supercomputers, largely by default. No foreign companies bothered to compete in the tiny market for the expensive and complicated machines, which can cost $10 to $15 million dollars apiece and require enormous sophistication on the part of the purchasers. The American computer colossus, IBM, dropped out of the field years ago, largely, some said, because the profits were too small, and the big company did not want to be accused of crushing all its smaller competitors. The two chief American manufacturers were Cray Research and a small offshoot of Control Data Corporation, joined recently by the even smaller entry, Denelcor.

Then the Japanese announced ambitious goals for a far faster supercomputer. But anxious Americans appeared to have overstated the Japanese prowess. The goal of the Japanese national supercomputer program has been repeatedly described in the United States as a machine, that, by the year 1989, would be a full thousand times faster than the most powerful existing American supercomputer, a truly prodigious leap ahead in such a short time. This "thousand times faster" goal had been cited repeatedly in the press, mentioned frequently by top industrial executives, and even appeared in official government reports circulated by the Office of Naval Research and, more important, by a panel of experts that prepared a report on "large scale computing in science and engineering" for the government two years ago.

But two key authors of that report would later acknowledge that they had been in error; the specifications announced by the Japanese were actually only 100 times faster than the most powerful American machine available when the Japanese program was launched four years ago. That would still represent a significant jump ahead, but it lacked the frightening ring of "a thousand times faster" and was, in fact, less than the 200-fold increase in speed that had been recommended in the United States by a federal coordinating committee. The coordinating comittee reported, correctly, that the Japanese were after a 100-fold increase.

Even more shocking than the ambitious goal announced by the Japanese was the revelation, early in 1984, that two Japanese manufacturers had already produced machines that were faster, on some standard performance tests, than the most powerful American machines. That, too, came to be seen as an overstatement. Measuring the speed of a supercomputer, it turns out, is far more difficult than determining the speed of a typist. The results depend heavily on what kind of problem the machine is given, how it is presented, and on whether the machine is rigged, or "optimized," to perform well on the particular test.

The experts had been arguing for months over what the tests on the Japanese machines proved. But one expert whose analysis is given great credence by federal officials — Jack Worlton, a fellow at the Los Alamos National Laboratory in New Mexico — said he believed the American machines were still faster. In a paper, Dr. Worlton concluded that the Japanese machines, which have only single processing units, were roughly comparable in execution rate to a Cray X-MP machine with a single processing unit. But since the Cray machine actually had two processors, the full array presumably remained significantly faster than its Japanese competitors. Dr. Worlton said he considered the Cray line of supercomputers to be "meaningfully ahead" of the Japanese, conceivably by factors of "two or four or something like that."

Supercomputers could be seen as unquestionably important in solving a variety of scientific and technical problems that require an enormous amount of computation, but they were not viewed as likely to change the world power balance. They were currently used to help design nuclear weapons, analyze geological data to search for oil, predict weather patterns, design highly complicated electronic circuits and chips, study nuclear fusions reactions, and even produce animated movies. Faster machines could solve a host of important problems that lay beyond the current computing capabilities, including the prediction of weather patterns in finer detail and the simulation of airflow around an entire airplane instead of simply around the wings, according to expert panels. But there would always be new problems to solve, many industry leaders said, and there would always be a demand for faster machines to solve them. It is largely a question of how much money and effort to expend on faster number-crunchers.

The military significance of supercomputers is limited. The defense and intelligence agencies use supercomputers to crack enemy codes, analyze weapons effects, and predict the weather, but overall, Robert S. Cooper, director of the DARPA, testified last year, "Defense has little interest in the current fastest computers available, that is, the supercomputers for number-crunching."

The economic importance of supercomputers has traditionally been minor. But industrial interest is slowly building, and some analysts believe the Japanese sensed a large untapped market for supercomputers. Indeed, the most surprising aspect of the new Japanese supercomputers is not their speed but the fact that they are "compatible" with IBM equipment and software programs whereas the American supercomputers are not. Thus, for the first time, businesses that used high-speed IBM computers would now be able to upgrade to a supercomputer without junking their old systems or learning a whole new system. If that began to happen, IBM would presumably be forced into the competition, and the nature of the supercomputer race would change. The big fear of American supercomputer enthusiasts is that the existing American manufacturers are overmatched by much bigger and better integrated Japanese companies. But those Japanese companies, in turn, are smaller than IBM. "If IBM gets interested," predicted Dr. Worlton, "they'll clean everybody's clock."

FIFTH GENERATION

Virtually all experts agreed that the drive to create radically new "fifth-generation" machines is potentially far more important than merely increasing the number-crunching power of conventional machines. The tantalizing promise of the "fifth generation" was that it could greatly extend the reach of computers in two directions — making them far more usable by ordinary people and allowing them to solve a whole new range of problems.

The core of the approach was that it would develop both machines and software programs that would handle not just numbers, the bread-and-butter of conventional computers, but also symbols, words, pictures and human speech in sophisticated ways that generally lie beyond the capabilities of existing numerical computers.

What's more, the new machines would use the insights of "artificial intelligence" — the discipline that had already produced computer programs that play chess and make some medical diagnoses — to perform rudimentary reasoning functions. The machines would apply a set of rules to an incorporated "knowledge base" in order to make simple choices, along the lines of "if you want the cheapest flight to Denver, then you'll have to leave on Tuesday" or "if the missiles appear in sector D, then the lasers are the best weapon to fire against them."

By most accounts, the Americans were way ahead in artificial intelligence research and related disciplines. But the Japanese stirred up the pot by announcing the first national program to organize the work and by setting ambitious goals. The fifth-generation project envisages machines that can accept spoken commands, translate from one language to another, interpret what its sensors "see," and provide expert advice on a range of topics, all by the early 1990s.

Japan's goals were economic and social; they focused on spreading the new machines into such areas as agriculture, fisheries, service industries, and small manufacturing operations that have thus far been little affected by computers, as well as into education, medicine and other social areas.

The chief American response came from the Defense Department's strategic computing initiative, which seeks to build a new array of intelligent weapons and management systems that can help offset the larger conventional forces of the Communist nations with ever more clever technologies. The goals for this decade include a driverless vehicle that can roar across the countryside, using visual sensors to dodge around barriers and machine intelligence to perform its battle mission; a computerized "pilot's assistant" that will respond to spoken language and help the harassed human pilot make split-second decisions in dogfights; and a battle management system that will help naval commanders ward off attacking missiles, bombs, torpedoes and other weapons.

On the more distant horizon lies an even more momentous military application: computer management of the proposed "Star Wars" defense against ballistic missile attacks. Any such system would have to detect and destroy in midair thousands of Soviet missiles and warheads in a matter of minutes should all-out war ever occur. That job simply cannot be done with existing computer power, defense officials say, and it may well require breakthroughs in artificial intelligence.

"The ultimate possibilities of artificial intelligence are truly mind-boggling and will some day come to pass," said Ralph Devries of the White House science office. But the field has a long history of false promises — computers have beaten the world backgammon champion but they still can't beat the world chess champion — and in recent years it has been in a state of stagnation, causing much handwringing among its intellectual leaders.

Most of of the major national and industrial programs in advanced computing that were launched with great fanfare have experienced severe start-up troubles, ranging from difficulty in getting the participants to cooperate with each other, a belated realization that impossible goals had been set, and lack of necessary funding. The Japanese fifth-generation program itself was denied full funding and was downgrading many of its projects.

Most experts said that the whole fifth-generation concept was so vaguely defined that it was difficult to discern where it was going or what its practical implications might be. Some even warned that the field was so unsettled that a crash program could be harmful, locking researchers prematurely into a handful of approaches and diverting scarce talent from the basic research still vitally needed to understand intelligence in both humans and machines.

MACHINE VISION: A HUMBLING LESSON

The great computer race was in fact one full of reason for caution and restraint. It could be, some computer specialists were saying, that very little intended to come out of this endeavor would actually materialize. In fact the whole quest was a bit humbling. After two decades of research into artificial intelligence, for instance, computer designers had yet to teach machines the seemingly simple act of being able to recognize everyday objects and to distinguish one from another. Vision, one of the seeming simplest of all human activities, seemed to be beyond the reach of computer builders. With relish their critics could dwell, point after point, on how little computer people actually knew about the human mind they were trying to mimic. They had, moreover, all developed a profound new respect for the sophistication of human sight and found themselves scouring such fields as mathematics, physics, biology and psychology for clues to help them achieve the goal of machine vision.

The result was a burst of new ideas, techniques and ways of building computers that suggest the riddle may be soluble after all. Dozens of universities and corporations have embarked on the quest with renewed vigor, including MIT, Stanford, Carnegie-Mellon, Columbia, Xerox, General Motors, General Electric and IBM.

Frustrating and rewarding at the same time, one recent insight of computer experts is that "what the mind wants to see" is often as important as what it actually "sees." Perception, in short, is a two-way street, a notion that has long fascinated poets and psychologists. "Things turned out to be much more difficult than we thought," said Dr. Tomaso Poggio, at the MIT Artificial Intelligence Laboratory. "It's clear why we underestimated the problem — it's so easy for humans to see." Almost any computer can be adapted to read bar codes in a grocery store or to "see" simple objects on an assembly line, according to computer scientists. But the nuances and grays of the real world are another story.

The difficulty is illustrated by the everyday accomplishments of human sight, according to the experts. Where the human eye sees a car, cat or other routine objects, a powerful computer peering through the lens of a television camera may see only disjointed jumbles of lines and circles. "So far no one has come up with a good general approach to the problem," said Dr. Takeo Kanade, an expert in machine vision at Carnegie-Mellon.

"People familiar with the physiology of the brain should have had suspicions," said Dr. John Kender, a computer expert at Columbia University. "By some estimates 60 percent of it is associated with vision."

Despite setbacks, the goal has increasing allure. Industry wants " vision machines" to aid robotic assembly, and the military wants them to help guide land vehicles and planes and to digest mountains of photographic data from spy satellites.

Two decades ago the field was filled with optimism, according to vision experts. They like to tell the story of how Marvin Minsky, a pioneer in artificial intelligence, once asked a graduate student to solve the problem of machine vision as a summer project.

ADVANCES MADE INITIALLY

Indeed, great strides were made at first. Computer experts in the 1960s made objects as visible as possible by using tricks of lighting and contrast to heighten their outlines. A camera would feed this information into a computer, which would analyze it for patterns that stood out — that is, for edges that fit together to make an image. The outline of this image would then be compared with ones stored in the computer's memory. A "horse" could thus be identified.

Even today, this principle is used by nearly all the hundred or so concerns that make industrial vision systems. Their computers look at high-contrast scenes, compare them with ones stored in memories, and decide if an object on an assembly line is damaged, is the wrong size, is in the right spot or whatever.

The problem is that conditions need to be almost perfect to make an identification, a fact that has kept eager vision vendors from selling as many machines as they would like. There is also the challenge of going beyond simple outlines. How does a computer separate "horse" from "rider" when their outlines merge? Worse yet, how does it identify "horse" when only the head of a horse is viewed by the television camera?

In contrast, it was easy to teach computers to perform feats of pure logic, such as playing chess. A bit harder was the computerized task of recognizing voices, a problem of perception that has certain similarities to vision but one in which greater progress has been made.

EARLY THEORIES DISCUSSED

According to Dr. Poggio, a bright and influential computer scientist who helped point out the complexities of the vision problem was the late Dr. David Marr of MIT. His posthumous book, *Vision,* published in 1982 by W. H. Freeman, is widely credited with generating much excitement among vision researchers.

In it, Dr. Marr pointed out that early in the century, Gestalt psychologists had noted the subtlety of human vision by showing that the perceived image was often greater than the sum of its parts. The brain, for instance, would find a pattern in a series of random dots or would see a whole circle, even if only parts of it were presented. In short, the brain often "saw" things that were not there.

In a similar way, poets have often said they use "visual inference" to convey a whole. "Nature red in tooth and claw," was the way Alfred, Lord Tennyson summed up nineteenth century discoveries about animal behavior that were at odds with prevailing ideas of universal harmony.

Dr. Marr said such phenomena hinted at a whole slew of assumptions that were "wired" into the visual system of the brain, and that did not bode well for the conventional approach to machine vision. Real vision would not be achieved, he said, by mere heightening of contrast.

Dr. Marr noted that dramatic lines in an image often came from such irrelevant factors as reflections or shadows, while many of the most important edges were nearly invisible. His solution was to create computer programs that first tried to identify surfaces, textures, colors and shadings. This step had to be taken, he said, before a machine could make the jump to identifying objects.

INSPIRED STUDY OF SURFACES

Today, working on their own or inspired by the writings of Dr. Marr, many vision researchers around the globe are pursuing the study of surfaces in order to enhance the vision of computers.

At Stanford University, Dr. Thomas O. Binford and associates use a basic cylinder shape as the starting point in a computer program from which they try to make identifications. A "human," for instance, might be made up of six cylinders: head, torso, two arms and two legs. So far, the group has had some success identifying airplanes from photographs taken at the San Francisco International Airport.

"It's given me great respect for biological vision systems," Dr. Binford said in a telephone interview. "Soon after we started this it became very clear to me that biological vision systems had great power and were very complex."

At Carnegie-Mellon, Dr. Kanade has created what he calls "origami world" as a way to simplify and model the complexities of the real world on a computer so it can make identifications. "It assumes the world is made of planar surfaces and that the operations you are allowed to do are cutting, folding and gluing along straight edges," he said. "At this point we're struggling for generality," he added, saying the point was to create general patterns for making identifications rather than specific ones, that is, using general surface contours rather than individual outlines of horses, cats or other objects.

At Columbia, Dr. Kender said vision researchers were also

starting to tackle the kinds of problems posed by Dr. Marr, riddles of inference, and how to identify objects whose parts are missing or indistinct.

VEHICLE THAT SEES

One hint of where the field is going, Dr. Kender said, can be seen from a recently announced goal of DARPA, which finances much of the advanced machine-vision research in the United States. In two years the agency would like to have the technology to build an "autonomous land vehicle" that would be able to use artificial eyes to follow a regular, well-defined road. The applications of such research are not all military.

Dr. Poggio said some of his recent mathematical work on image identification by computers had led biologists to look for similar processes in the brains of monkeys. The search in biology for clues to the problem of computer vision has begun to drive home not only similarities between man and machine but also profound differences. "The irony is that the brain is beating out computers with neurons that operate about a million times slower than silicon," Dr. M. Mitchell Waldrop wrote recently in *Science* magazine. "The secret, of course, is in the wiring: The neurons are in there doing millions or billions of operations simultaneously. Whereas computers, with a few exceptions, are still based on a serial, one-step-at-a-time architecture." Such insights into the complex "wiring" of the brain have dawned only recently.

Two decades ago, biologists were achieving breakthroughs in describing the anatomy and physiology of the eye and brain, and it was assumed that the "hows" of vision would quickly become self-evident. They didn't. What became clear was the astonishing diversity of neural connections and cells in the retina of the eye, in the optic nerve and in the visual cortex of the brain. Processing was going on all over, even in the eye itself. Yet how images formed remained a mystery. "They thought they only had to describe the biology," said Dr. Poggio. "But the real problem is how to explain what is happening and how things interact — to explain the brain's program." The upshot is that many of the "hows" of human vision remain a mystery. Nevertheless, some secrets teased from the brain are being applied by machine-vision researchers. At Columbia, Dr. David Shaw, the computer expert, has been working on vision problems and experimenting with parallel processing as a way to achieve the kind of blinding speeds of processing at work in the human brain. "What we've got is dramatically better than using a sequential computer," he said. "We're able to process some problems a thousand times faster than general-purpose scientific computers."

He noted, however, that so far the new architectures had not opened up any dramatic new paths and that overall solutions to the problem of machine vision could well be far off. "If we knew how to build brains we would," he said. "They're still the best there is for a lot of applications."

THE CRITICS SPEAK

Meanwhile, the U.S. computer effort was coming under criticism. The Defense Department plan to spend $600 million over five years to develop a new generation of computer-based military systems became the focus of a debate among a number of senior computer scientists. Carrying out the plan, according to its sponsor, DARPA, would "provide the United States with important new methods of defense against massed forces in the future."

The criticism was varied. Some scientists objected to the immediate goals of the project itself, contending that the aims could not be achieved and, indeed, that the project would substantially increase the chance of war. Others argued that, with the increasing role of computers in virtually every aspect of life, extensive military involvment in the development of computer systems posed a threat to important social values. These critics contrasted the military aims of the Strategic Computer Initiative, the largest single U.S. research effort in the computer area, with the goals of the fifth-generation project in Japan. The Japanese project was explicitly framed in terms of enhancing productivity and improving social services such as schools and medical care. The Pentagon project envisioned the development of machines with humanlike skills that would be integrated into specific military applications: smart tanks, robotic jets and mechanical consultants in sea battles.

Jan Bodanyii, a Defense Department spokesman, defended the thrust. "The Strategic Computer Initiative will not skew American values," he said. "It simply is a project to look at certain kinds of advanced technologies to see if they can be used for military purposes." Michael L. Dertouzos, director of the Laboratory for Computer Science at MIT, said that "the idea that the source of funding influences and controls the ultimate deployment of a discovery is just plain false."

The most recent criticism of the computer initiative came from the Computer Professionals for Social Responsibility, a nonprofit educational group founded in the early eighties that had more than 500 members in chapters in ten cities. In a formal statement, the group compared the current Pentagon computer project with the drive to develop nuclear weapons in the 1940s. "Past attempts to achieve military superiority by developing new technology, rather than increasing our security, have brought us to the present untenable situation," the group said. "The push to develop so-called intelligent weapons as a way out of that situation is another futile attempt to find technological solutions for what is, and will remain, a profoundly human political problem."

Another critic was Mark Stefick, a scientist at the Xerox Palo Alto Research Center in California.

"The American plan is rationalized by military needs, and secondarily, commercial values," he wrote in an analysis. "The Japanese plan is rationalized by social needs, and secondarily, commercial values. This apparent contrast of social versus military orientation has raised concerns even among those who concede that DARPA is the only American institution with the scale of funding and vision of the research community to do it."

In Japan there was criticism, too. Since setting up shop in June of 1982, Fuchi's group was faced with critics who claimed that the fifth generation was largely a futuristic dream with a lot of public relations appeal. And there were budgetary problems, as cutbacks in funding started to take away some of the drama and momentum established earlier.

But there were those whose optimism remained strong. "They will get a lot of payoff even if they don't reach all their targets," said David Brandin, a vice president of SRI International in Menlo Park, California, and the past president of the Association for Computing Machinery. One possible payoff that he suggests: a leg up on making computers easier for people to use. For his part, Fuchi said he doubted that his group would have a fifth-generation machine ready for the marketplace when the project is completed. "But if we can develop basic technologies that will lead to a new age of computers," he said, " the project will be a success."

Meanwhile, the United States was plugging along. Its computer engineers and software designers were continuing to develop early versions of artificial intelligence, the "expert systems" that would help in making medical decisions or solving problems in automobile mechanics. They were continuing, with considerable vigor, to pursue the parallel processing technology that would enable progress in a fifth generation. In direct response to the Japanese, a private American consortium, with nineteen member corporations, called the Microelectronics and Computer Corporation, was begun in 1983. It had much in common with the Japanese computer effort but, if anything, its agenda was broader. MCC set up a ten-year program in advanced computer designs that included under its umbrella data-base management, human-factors technology, parallel processing and artificial intelligence. Although it took a couple of years for the corporation to get its act together, it came to include more than 400 employees and an impressive laboratory hard by the University of Texas campus at Austin. Also, at least partly in response to the Japanese push, DARPA announced its Strategic Computing Program. The agency, by 1987, had already spent $30 million on parallel processing alone, and it was expected that DARPA might spend as much as $100 million.

At the same time, Star Wars, President Reagan's plan for an antiballistic missile shield that would protect the entire nation from attack by the Soviet Union, was a powerful inspiration to the artificial intelligence field and to supercomputing in general. Whether you liked it or not, it was clear that major technology development in the United States was often quickest and most far-reaching when it had some big defense plan underpinning it. Star Wars, which entailed the development of a system that might have to track thousands of warheads at a time, could not work without magnificent new supercomputers.

And supercomputing was central to more specific defense concerns, too. For instance, there was this oft-discussed military problem: given the position of enemy radar and a destination, what flight path should a pilot take to minimize the chance of being detected? It was a problem that was often faced, as in the bombing of Libya, and it would take tremendous computing power to resolve it efficiently. In addition, more than 100 projects to build advanced computers had been initiated by American companies and universities. Among them were the University of Illinois, the California Institute of Technology and MIT. IBM alone was spending tens of millions of dollars on parallel processing.

The United States, in the second half of the decade, had recovered some of its poise and confidence following the Japanese shock. Its supercomputing and artificial intelligence efforts looked to be in rather good shape and not at all suffering by comparison with the Japanese.

In Japan, the yen was in some trouble. There were budget cutbacks in the supercomputing efforts, and the results of those efforts, in any case, were only modest. The fifth generation was now seen to be the project of human beings rather than omnipotent engineering giants. Not only was the fifth-generation aspect of the Japanese effort being cut back in budget, scale and perhaps aspirations, but the less ambitious Superspeed Project was in considerable difficulty. Well into the second half of the decade, the Japanese had not developed a single prototype.

"All of a sudden, it doesn't look so threatening," said one American supercomputer executive.

A fascinating new development had burst onto the stage in America at the same time. Rapid-fire advances in achieving superconductivity — the ability to conduct electricity without the loss of any energy — were being immediately funneled into computer applications. The Japanese were watching closely and pronounced themselves ready to make whatever use they could of the new technology.

In any event, while there was still much talk of competition in the air, the American, European and Japanese computer designers so determined to develop another generation were talking a lot about cooperation too. It was beginning to look as if all these people really did need each other.

Chapter Three

The Race for Supercomputers

By Peter H. Lewis

Peter H. Lewis is a computer columnist for The New York Times.

The most powerful computer in the world stood in the center of the testing room, its twelve sleek, vertical columns arranged in a tight semicircle like a futuristic Stonehenge.

Along the walls, a row of boxlike mainframe computers, each capable of performing the work of tens of thousands of human clerks, waited for the instant when the machine would flash to life in a frenzy of calculations so furious that its circuitry would melt without constant refrigeration.

The mainframes, which only a few years earlier had themselves been thought to be among the world's most powerful computers, now served only as attendants for the starkly beautiful Cray supercomputer. They were needed to receive and process the torrents of data that would spew forth, 100 million bits a second, from the new machine.

When it was introduced in 1976, the Cray 1 inherited the title supercomputer, a name given to any computational engine that is the fastest and most powerful at the time. It was a staggering four times faster than the computer it eclipsed, the Control Data CDC7600, and heralded a new age of large-scale computing — one based on the then-revolutionary semiconductor material called silicon.

Less than twenty years earlier, one of the first supercomputers, the Illiac-I, a lumbering giant with flickering tubes and clicking switches, dazzled the scientific community by finding the largest known prime number in 100 hours, surpassing centuries of work by mankind's most brilliant mathematicians.

In the test room at the Cray Research Development Facility in Chippewa Falls, Wisconsin, a technician fed the prime number algorithm to the new Cray 1. Ten seconds later the Cray 1 was done, patiently waiting for a more challenging job.

Now a decade old, the Cray 1 remains the standard against which supercomputers are measured — only now it is the lowest denominator, not the highest. And the silicon technology that allowed the Cray 1 to do in ten seconds what had taken Illiac 100 hours has been pushed nearly to the limit, constrained at last by the slowness of light and the bulk of electrons.

Supercomputers have reached a crucial point in their evolution. Already, computer engineers are exploring new designs to meet the increasingly complex scientific problems of the next decade, problems that require trillions of calculations for such tasks as nuclear weapon design, graphics, meteorology, geophysics, oceanography, aircraft design, oil exploration, robotics, satellite imaging and even the design of new supercomputers.

Supercomputers that can perform more than a billion calculations a second, including the second-generation Cray 2, are now being installed around the world. The Cray 3, scheduled to be completed in 1989, is said to be several times that fast. And yet another, unnamed, Cray supercomputer, set for completion in 1992, has as its theoretical goal a speed several times faster yet.

Still, government and scientific researchers say, it is simply not fast enough. The solutions to many problems being faced today, including some that the American military considers vital to the country's security, are beyond the reach of any supercomputer now on the drawing board. As soon as the supercomputers arrive at the solution to one problem, they seem to create a dozen other questions that are vastly more difficult.

A race is now on between the two great computer powers, the United States and Japan, to produce a supercomputer that will take scientists into the twenty-first century. The contenders in the commercial market are Cray Research Inc. and Control Data Corporation's ETA Systems Inc. in the United States, and three Japanese companies, Fujitsu Ltd., Hitachi and NEC, formerly Nippon Electric Company.

Two trends in supercomputing are now apparent. The first is to increase the speed, memory and power of single processors. The second is to create a radically new type of computing engine called a massively parallel system — tens, hundreds or even thousands of processors strung together to harness their collective power, making a machine far greater than its own parts.

By combining these two streams of research, supercomputer companies believe they can develop machines in the 1990s that can handle an almost inconceivable 100 billion floating point operations — complex additions or multiplications — in one second. The U.S. and Japanese governments have vowed to spend a billion dollars each in the effort.

If they succeed, and history suggests that they will, every imaginable scientific realm will explode with discoveries. But more exciting is the possibility that these supercomputers will discover new realms beyond the imagination.

"We're at a pivotal point in computer history," said Dr. Salvatore Stolfo, a computer scientist at Columbia. "Fifty years from now people will look back on this transition with almost mystical respect."

In the meantime, supercomputers are already having a profound influence on research, even in fields that are not traditionally computer dependent, like political science, economics and archeology.

"The supercomputer is the single greatest impact on world communication, automated factories, health care delivery, biotechnology production, renewal of basic industry and heightened productivity of the service industry, including government," said

George Kometsky, a professor of management and computer sciences at the University of Texas at Austin.

THE IMPACT OF SUPERCOMPUTING

The impact of the scientific supercomputer, unlike the so-called fifth-generation machines that are being designed for artificial intelligence, comes from the sheer brute force of its computational speed, its vast memory and its great precision. Such power is revolutionary in science, which has advanced for centuries with just two tools, the brain, for theory, and the hands, for experimentation. For the first time, problems that were unsolvable now can be solved.

What problems are so difficult that they will occupy hundreds of these superelectronic brains for years to come? In part, the problems are a direct result of previous advances by computers, leading to an ever-accelerating cycle of discovery and quest.

As computers give scientists ever-greater knowledge of the workings of the natural world, the scientists are able to construct increasingly detailed models of phenomena that cannot be tested in the laboratory at reasonable safety or cost, or even tested at all. One of the leading uses of current supercomputers, for example, is in the design, modeling and simulated testing of nuclear weapons.

Another example is the study of aircraft aerodynamics, testing the turbulent flow of air over various surface shapes at different speeds and under a variety of weather conditions. Because any given area on the planet's surface includes an almost infinite number of points to observe, only a supercomputer, with its vast memory and speed, can do the job in a reasonable amount of time. The wings of the Boeing 767 aircraft were designed and tested in just a few weeks by engineers using a $15 million supercomputer. Previously, the process would have required thousands of man-hours and many months of research in huge wind tunnels that cost $150 million a year to operate.

Nevertheless, the modeling of turbulent airflow around an entire airplane, especially one as complex as the proposed "Orient Express" craft that can traverse atmosphere at supersonic speeds and cross the boundaries of space, is beyond the capabilities of current supercomputers by a factor of 100 or more. Computers as we have known them for the past forty years simply cannot be pressed that far, computer engineers say.

The barrier that designers of the new breed of supercomputers must contend with is the speed of electricity. Under the best circumstances, electrons coursing through a microelectric circuit travel ten inches in one billionth of a second, or nanosecond. Because this speed is essentially finite and cannot be altered, engineers from the beginning have sought to make computers go faster by cramming more circuits and wires into ever smaller spaces, thereby reducing the distance the electrons must travel.

SMALL IS BEAUTIFUL

In this sense today's machines are a far cry from the first computers in the 1940s, which relied on clumsy electromechanical relays, and those of the early 1950s, when bulky, capricious and power-hungry vacuum tubes made the machines as big as houses and required small armies of white-coated technicians to replace overheated tubes. With the development of transistors in the 1950s, a second generation of machines, smaller and more reliable, came on the scene.

Also in the 1950s, engineers turned their attention to silicon, a semiconductor named because of its place on the periodic table between the conductive metals and the nonmetals. It is just below carbon, and, in fact, silicon crystals can be cleaved like diamonds, giving them the electrical properties needed in the new field of microelectric circuitry.

Unlike other semiconductors, the researchers discovered, silicon formed a surface layer of insulating silicon dioxide — glass, essentially — when exposed to steam. These two properties, conductivity and insulation, on the same substrate quickly attracted the attention of computer engineers. Several transistors, capacitors and resistors were plastered to a crude wafer of silicon, and the integrated circuit was born.

Over the years this so-called small-scale integration grew larger, with hundreds of transistors and other parts on each chip, and larger still, to what is referred to today as very large-scale integration, with as many as 2 million components on a chip. There are almost half a million parts on the tiny memory chips used in most supercomputers today.

There are two main advantages to such densely packed circuitry. First, it allows designers to construct the large memory units that supercomputers require to store data, and second, it significantly reduces the distance electrical impulses must travel to complete a cycle of operation.

Some companies are now experimenting with ultra-large-scale integrated chips containing millions of components. According to J. Jeffrey La Vell, a technological strategist for Motorola, by the end of the decade there could be chips on the market with 20 million components, a feat he described as similar to "designing, mapping and monitoring the highway system for metropolitan Los Angeles in an area one-twentieth the size of a postage stamp."

But some supercomputers are made of hundreds of thousands of such chips, and even the short wires connecting them can add miles to the distance the electrons must traverse. Scientists at several companies are exploring the possibility of eliminating such connecting wires by, in essence, increasing the size of the postage stamp.

THE WAFER: A MONSTER CHIP

The construction of the tiny computer chips used today begins with a wafer of highly polished silicon about five inches in diameter, which in a series of steps called photolithography is etched with about 200 "chip" patterns. A light-sensitive emulsion is spread on the wafer, exposed to light through a mask, much like a photographic negative, and then "developed" to create the circuitry. After testing, the wafer is broken into its individual chips, the chips are encased in carriers, the carriers are attached to printed circuit boards and the boards are wired together to form the computer.

"You chop the wafer apart and then the first thing you do is put it back together," explained Dr. Robert R. Johnson, president of Mosaic Systems Inc. of Troy, Michigan. "It's sort of nutty, but that's how the industry grew up."

By covering the entire surface of a wafer with integrated circuitry, the wiring that now connects hundreds of individual chips could be eliminated, and thus the speed could be increased by several magnitudes. The drawback is that a tiny flaw anywhere in the millions of microscopic circuits would make the wafer worthless, so precise etching, testing and correction of circuits, already a costly and time-consuming task, would become commensurately more difficult.

The solutions to such microelectronic nightmares are being sought through laser-induced chemical reactions, which avoid the traditional method of etching circuit designs on chips by photolithography. Such research into "laser pantography" is being done at MIT's Lincoln Laboratory, Columbia University and the Lawrence Livermore National Laboratory in California, among other places.

At Livermore, tiny pulses of highly focused argon-ion laser light are beamed at the wafer through a cloud of gas. As the laser moves across the wafer, through different types of gas, circuit lines less than a millionth of a meter wide, with varying conductive properties, can be etched on the silicon. This relatively simple method of "direct writing" on a chip eliminates several steps in the manufacture of wafers, and also holds great promise in repairing, modifying and monitoring circuits during the etching process.

Stacks of these "monster" wafers could then be strung together to make even faster new supercomputers.

But there are major drawbacks to packing circuits so tightly. The main one is heat. Supercomputers must have some way to keep their figurative blazing speed from becoming literal, reducing their internal circuitry to a molten puddle of wires and glass. Fans, which are adequate to cool home computers, are far from adequate for the highly compressed innards of supercomputers, and even ingenious internal refrigeration systems are giving way to new cooling technology. Some supercomputer processors are now suspended in tanks of liquid helium or fluorocarbon.

Another drawback is that silicon itself has limitations, and researchers are now turning to other semiconductors. The next generation of Cray supercomputers will use the compound semiconductor gallium arsenide (GaAs), instead of silicon. Another candidate is indium antimonide.

The same density circuitry on a GaAs wafer can perform perhaps five times faster than on silicon while using less energy, the researchers believe, while also providing superior insulation and thus lower heat. Gallium arsenide chips also have a higher tolerance against radiation, making them ideal for use in complex electrical systems in space, or in weapons. GaAs components are already being used in some microwave communication systems where silicon circuits are too slow, but GaAs wafers are expensive, about ten times the cost of comparable pieces of silicon.

Even so, some researchers believe that the upper speed limit of the fundamental logic functions of computer circuits — the switching from one electrical state to another, open and shut, thus acting as a gate to direct the flow of current — is 3 or 4 billion times a second under optimum conditions.

PARALLEL PROCESSING

How, then, do computer designers plan to reach 10 billion, let alone 100 billion, operations a second within a decade?

The leading technology today is parallel processing, that is, arranging many individual computers in such a way that they work concurrently on one problem, much the same way as thousands of neurons in a human brain attack a problem from different perspectives at once. The emphasis in computer design is shifting from the so-called serial, or von Neumann, processors — named after the American mathematician who established the system in which numbers are processed in sequential, assembly-line fashion — to multiprocessors arranged in parallel architecture. Depending on how many processors are strung together, and how efficiently they can be made to work in tandem, these parallel systems can perform thousands or perhaps millions of functions concurrently.

For example, the Cray 1 may be 1,000 times faster than the IBM XT personal computer, but researchers have found that 100 XT's patched together in an effective parallel network have the same computing capacity as the Cray at a fraction of the cost. By replacing the XT's in the parallel network with supercomputer processors, the speed, in theory, could be increased more than a thousandfold.

"The parallel processor is a more natural kind of machine," said Paul Castleman, president and chief executive of Bolt, Beranek & Newman Advanced Computers Inc. His company takes more "mundane" technology — in this case the mass-produced 68020 microprocessors used in the Apple Macintosh II personal computer — and strings scores of them together to achieve power that exceeds that of the original Cray 1 and costs much less.

That is also the philosophy behind the series of Cray super-computers designated X-MP, for experimental multiprocessor. The X-MP's have two or four central processing units, each more potent than that of a single Cray 1, giving them many times the power. Yet the X-MP's cost as little as $2.5 million, as against $12 million to $15 million for the original Cray 1.

The Cray 2, introduced this year, has two or four "engines" with twelve times the power of the original. And within a year, an X-MP with eight processors is expected to deliver four times the power of the Cray 2.

Nonetheless, according to John Rollwagen, chairman of Cray Research, "using supercomputers is much harder than making them." This is especially so in parallel processing.

The success of parallel systems depends on how well the problem to be solved can be broken down into smaller segments, or granules, that can be divided among the individual processors. If a problem does not have good "granularity," the hardware will run inefficiently. On the other hand, trying to force granularity into a problem may make solving the problem inefficient.

For example, "dance hall" parallel processors have memory units in a row on one side and processors in a row on the other. To keep everybody "dancing" all the time requires more advances in algorithms, programming languages and support software.

"It is a little bit like having Democrats on one side and Republicans on the other at budget time, and we hope they can hand shake and do everything that is necessary to give us the performance that is necessary," Dr. Edith W. Martin, former

Deputy Undersecretary of Defense for Research and Advanced Technology and a former executive with the Control Data Corporation, explained to a congressional panel.

Software, the instructions that make the hardware do what it is supposed to do, is absolutely critical to the effective performance of the new parallel supercomputers. It is also an area that has been the weak link in supercomputer technology for the past decade. "If we are going to exploit our hardware advances," Dr. Martin said, "then they must be accompanied hand-in-glove with comparable advances in software capability. Software cannot be a stepchild."

The problem has been to find programmers familiar with supercomputers. It is believed that about half of the 230 or so supercomputers now in operation are used by the military, which limits access to the machines. Another limiting factor is the high cost of supercomputers; even at the comparatively bargain price of $2.5 million, they are prohibitively expensive for most universities and businesses.

Because so few new programmers have experience with supercomputers, those who do can command high salaries. The development of software is believed to account for as much as 90 percent of the cost of maintaining a supercomputer.

In the past, supercomputers have been delivered without software, so only those companies or institutions with highly trained staffs of programmers could use them. Now, however, companies are discovering that they can recoup some of the development costs by licensing software to other users. About 400 such programs for the Cray are available.

Still, parallel processing presents a particular problem for software developers. "Development of future generations of supercomputers based on parallel computing is a systems problem, not merely a problem in developing computing hardware," noted Dr. J. C. Browne of the University of Texas at Austin.

THE JAPANESE CHALLENGE

Thus, it came as a shock to American computer designers when they learned that Japanese researchers were developing supercomputers that were just as fast as any machines the American had on the drawing board, and, more astonishingly, could use off-the-shelf software developed for IBM System 370 computers, a popular mainframe system in American businesses.

According to Dr. Steven A. Orszag of MIT, the Fujitsu VP-200 took "the best features of Cray, CDC and IBM architecture and put them all together.

"What's revolutionary is that these fast Japanese machines could be used by business or government," Dr. Orszag said. "In the past, the uses of supercomputers have been much more specialized and mainly scientific."

After years of unchallenged dominance in the development of supercomputers, the United States suddenly found itself in an expensive, high-stakes duel with Japan for supremacy in the field.

Although they were aware that the Japanese government had embarked on an ambitious program to aid its computer industry, American researchers say they were stunned by the speed, depth and breadth of the Japanese advances. In 1987, for the first time, a Japanese supercomputer, made by NEC, was sold to a research consortium in the United States. It was the first time Cray had lost a bid in its own country.

The Japanese program was designed with the specific goal of developing the world's fastest scientific supercomputers, at least 100 times more powerful than anything existing at the time, as well as a new generation of computers — a fifth generation capable of artificial intelligence. Directed by the Ministry of International Technology and Industry, it is a concerted effort involving the government, industries and universities.

The United States, at first stunned by the significance of the Japanese effort, eventually moved to match its rival. In February 1985, the National Science Foundation, citing a "desperate, overwhelming need" for access to supercomputers by academicians, pledged $200 million to establish supercomputer centers at four of the country's leading universities and asked state governments and private industry to match the outlay.

The emphasis to get supercomputers into the schools was a significant departure for the United States, which had concentrated its supercomputer programs in the military. In fact, while the military has been using the most advanced computers since the late 1940s, it was not until 1982 that the first supercomputer was operating at an American university.

By concentrating supercomputer research in the universities, the National Science Foundation had three goals: to improve the quality of basic research, since even now there is poor access to supercomputers outside the military; to train students for a time when computers will undoubtedly play an even larger role than they do today; and to increase the market for the country's supercomputer manufacturers.

Venture capitalists, the specialized firms that raise money for new technological adventures, are extremely reluctant to invest in an expensive, long-term development project in an industry where machines are born and become largely obsolete in just two years, and where the total market may be a few hundred machines at best. And though it takes at least two years to build a new supercomputer, it takes ten years to produce a computer specialist. At present, schools in the United States are turning out fewer than 300 new doctorates in computer science each year.

"Computer science is at a pivotal juncture," said one of the new specialists, Dr. W. Daniel Hillis of the Thinking Machines Corporation, a spinoff of MIT. "Nobody knows exactly what they're doing in the parallel area. But that's the nature of pure research. It's an incredibly exciting time to be in the field. Things that we only dreamed about are suddenly becoming possible."

THE TWENTY-FIRST CENTURY

Parallel processing is one of the dreams that is becoming possible today, but other supercomputer technologies are still in the dreaming stage. Some of the technologies that are being explored for the twenty-first century are novel, and others are seemingly bizarre.

Because the computer is so closely identified with electronics, it is easy to assume that it must be electronic itself. Nonetheless, other formats exist, including computers that use photons instead of electrons and computers based on genetically engineered bacteria.

Experiments with superlattices, in effect thousands of layers of crystals stacked atop each other yet equaling the thickness of a sheet of paper, may have promise for applications in so-called optical computers.

With lasers, these computers could raise the operational speed of circuits to picoseconds, or thousandths of billionths of a second. In other words, according to theory, the optical logic gates could carry out a thousand billion operations a second.

Also, optical switches conceivably could have more than two logic states, the on and off of current semiconductor switches. And unlike electrical currents, which would become mixed in a single circuit, several laser paths could go through a single crystal and remain separate, so the original signal could be subjected to a handful of logical operations at one time.

A drawback to optical computers is the low operating temperature needed to offset the heating caused by high energy lasers. Also, present laser technology does not permit the very fine tuning of lasers on such small scales.

But working optical technology is, in the relative time frame of computers, a long way off. Even more remote are organic computers made of single molecules that exist in one of two states, on or off, using the switch analogy. Such molecules are now being synthesized in laboratories, but to date no one has been able to figure out how to string them together with tiny little wires to form a computer. These molecular switches, if they can ever be harnessed, could be attached to bacteria that had been genetically altered to form specific sequences of proteins. By enticing the switches to attach themselves to certain proteins along the chain, organic circuits could be formed.

The scientific community may be waiting for a long time to see these fanciful new technologies, but it is not waiting to develop new applications for supercomputers. "Supercomputers will become the wind tunnels, test tracks, materials laboratories and chemical laboratories of the future," said Peter A. Gregory of Cray Research.

Chemists are using supercomputers to "slow down" chemical

reactions like combustion, to test and refine experimental drugs on electronic patients, and to store databases for DNA and RNA sequences.

Geologists are studying the flows and effects of pollutants in underground aquifers and studying the erosion of coastlines. Through seismic monitoring, oil companies are determining with great precision the location of oil deposits and learning how to extract the most oil from existing reservoirs.

Economists are making detailed nationwide studies of inflation, unemployment and economic growth, testing policies and weighing social costs.

Meteorologists are refining their short-term and long-term weather forecasts, with significant implications for agriculture and predicting natural disasters.

Industries are using supercomputers to developing robot vision and control, machine control, quality monitoring and scheduling.

Automobile makers are redesigning cars and engines without actually building a single prototype, gaining greater efficiency and fuel economy. By "crashing" their designs on a computer screen, they are learning to make safer cars.

Doctors are using the graphics capabilities of supercomputers along with CAT scanners to generate three-dimensional Xrays. Cancer researchers are studying the innermost secrets of carcinogens and making maps of human genes.

Filmmakers are generating dazzling and realistic graphics images for motion pictures. But one second of computer-generated film in a science fiction movie requires billions of calculations to produce realistic animation, a time-consuming process even with the most powerful Cray now on the market. In the science fiction script of tomorrow, supercomputers will be able to generate such images in "real time," allowing fantasy to unfold before the viewer's eyes.

And when the fantasies of today's supercomputer designers actually take shape in the machines of tomorrow, the scientific community will be waiting. In the words of one researcher, "Whenever somebody thinks of a faster computer, someone comes up with a way to use it."

Chapter Four

The Biochip: Merging Computers and Biology

By Michael Edelhart

Michael Edelhart is an editor of P.C. Week, *a newspaper about corporate use of computer systems; he has written several science books, including* Assembly Language Primer for 80386 Computers.

The future may not be etched in silicon.

Tomorrow's computer chips probably won't be built from inorganic metals laid in thin strips on purified silicon beds, as they are today. They may not even carry electrical current. Instead, so-called biochips and molecular chips will draw their circuitry designs from the complex process of nature; bits of data will be stored in organic compounds and manipulated by light or the mixing and matching of molecular chemistry.

What computer architects hope to accomplish with these silicon chip alternatives is the development of circuits that can out-micro the microchip, run programs like a Ferrari compared with today's horse and buggy and help give desktop micros the intelligence to do several jobs simultaneously, like the human brain. Ultimately, this next microchip revolution may even merge electrical engineering and genetic engineering. Some scientists predict gene-splicers may someday invent bacteria that will generate specially designed molecular composites for computer use.

This intricate connection between physical and biological sciences has even excited some visionaries to predict the emergence of molecular computers that design and reproduce themselves according to DNA blueprints and that can be implanted in the human body as an integral part of a person's ability to control his senses and manipulate information from the outside world.

The biochip revolution today is far from reaching these lofty aims, however. Right now, it can be found in a few small, but growing, yet often unrelated cells of researchers at universities and research parks scattered around the globe. Some of these researchers are computer specialists, others geneticists, a few materials engineers. Their projects range from tinkering with nerve cells to electrifying polymers. The sole factor these scientists have in common is that they are entranced by the idea of ultimate miniaturization of information, of which the tight, incredibly informative codes of nature are the apotheosis. And so they are working on new techniques for coding and processing information within individual man-made molecules in ways vaguely similar to natural processes. This quest, they hope, may someday open the way for microscopic, three-dimensional biochips that make today's inorganic microprocessors seem like slow and ancient mammoths. Interest in organic and biological chips has hovered around the edges of the computer business for more than a decade, fueled by a realization that silicon-based components are rapidly approaching the limits of their size and speed. High-density semiconductor chips today cram some 450,000 transistors into a 4 millimeter square, with about 1.5 microns separating each component. Even the most advanced techniques on the horizon today could only shrink that spacing down to .2 microns, four times better than current capabilities. Beyond that level, the technology falls prey to reliability problems and bumps up against the rough edge of atomic physics, where ordinary rules of performance no longer hold.

Faced with this implacable reality, some scientists began

thinking about taking computing inside the molecule. At molecular size, computer circuit elements could operate at thousands of times the speed of today's chips and could be squashed much closer together, since molecular processes produce little heat.

Another advantage of molecular microprocessors is that they would be more resistant to the electromagnetic pulse effect of nuclear explosions, as well as to more conventional "crosstalk," interference between circuits.

All the way back in 1974, Arieh Aviram and Philip Seiden of IBM and Mark Ratner, who now teaches at Northwestern University, proposed using carbon-based molecules as computer switches. Many organic compounds, they noted, exist in two different stable electrochemical states. Specifically, they pointed to a group of chemicals involved in biological electron transfer known as hemiquinones.

The movement of a hydrogen bond from one part of a hemiquinone molecule to another shifts the chemical between its two stable electrical states. Voltage applied across the molecule can cause it to jump from one state to the other. In theory, such a compound could serve as a computer switch.

Aviram and Seiden actually took their idea far enough that they patented a molecular switch designed around an electron donor compound (TTF, tetrathiofulvalene), and an electron accepting compound (TCNQ, tetracyanoquinodimethane) with an insulating layer between them. When the molecular sandwich is placed between electrodes and a high voltage is applied, a quantum-mechanical phenomenon called tunneling occurs, and electrons jump, paradoxically, from the acceptor molecule to the donor, creating a stored electrical charge and making the sandwich a switch. Robert Metzger, now of the University of Alabama, and Charles Panetta, a chemistry professor at the University of Mississippi, Oxford, managed to synthesize the compound TTF-TCNQ, but have found it difficult to purify and troublesome to work with, so little has been done to test its switching capabilities.

More recently, Metzger has been trying to determine the best method of bridging the two compounds. The likely candidate is some sort of acid or alcohol derivative. "The trick," Metzger says, "is to build the bridge before the ends meet" — referring to the intense attraction the compounds would have for each other when holding opposite electrical charges. Once a bridge medium is found, the researchers have to figure out how to lay down a layer of compound one molecule thick and how to put the tiny construction between two pieces of metal to create a diode switch. It's a daunting task and not even the research team is certain of success. "I've got some doubts," Metzger states. "One is whether the molecule can be made. Another is whether it's going to work."

Eventually, though, the team hopes to place the molecules on thin films and build a switch much like the IBM pair's patent.

The practical problems encountered by the Alabama–Mississippi researchers, and others, have forced the high end work in molecular switch studies onto a purely conceptual plane. The leader in framing the possibilities for molecular electronics is Forrest L. Carter, a chemist at the Naval Research Laboratory in Washington, D.C. He has suggested using electrically conducting polymers, such as polysulfur nitride or certain versions of polyacetylene, as the molecular wires for a biologically based chip. These compounds have peculiar and not yet well studied electrical properties that have been linked to "solitons," wavelike disturbances that sweep along the polymer chains and result in electrical conduction.

The theory of soliton conductivity is highly controversial and highly speculative. It is based upon the idea that in long, stringlike compounds with variable single and double valence bonds, electricity can result from a tsunami-like ripple along the spine of alternating bonds. Transpolyacetylene, for example, is a drunken ladder of carbon groups linked with alternating double and single bonds. Solitons are thought to sweep along the compound chain at nearly the speed of sound, cresting at about every fifteen carbon-carbon bonds. The soliton flow displaces the bonds and causes an electron flow.

Experimental evidence of soliton effects remains scanty today. Some suggestions have been found in experiments with crystalline arrays. But no significant indications of the transpolyacetylene effect has yet been found in the lab. To produce his conceptual models of what soliton-based switches might be like,

Carter simply assumed the accuracy of the theory and then built upon it.

In Carter's conceptual framework, light-sensitive molecules called chromophores embedded in a conductive polymer like transpolyacetylene could set off a soliton reaction. Creating a soliton wave along the compound would rearrange the chemical bonds of the polymer in a regular pattern. This shift in the bonds would drag the chromophores along for the ride and would result in a shift in the chemical's responsiveness to a strong light source. Elimination of the soliton would return the molecule to its original state. Thus an on-off switch could be made from a compound, based upon soliton flows.

A more complex Carter model envisions three soliton channels combining to activate a switch. A pair of polymer chains would be linked to a single chromophore. Soliton waves down the two channels would "activate" the light sensitive compound. Then, the addition of electricity from a third soliton channel would cause the chromophore to emit a photon of light, signaling that it was "on." The original activated state would represent "off."

Carter has even gone so far as to postulate that solitons will flow only along chains where single and double carbon bonds alternate. By placing pairs of single or double bonds back to back, he feels, a designer could direct or block soliton flow. This, in his opinion, points toward the possibility of creating molecular logic and memory circuits where the flow of data is controlled by the arrangement of the chemical bonds.

He calls such circuits soliton valves. In one design, Carter proposes placing a single carbon atom at the juncture of three soliton channels. If a soliton is generated down one chain, it would sweep around the central carbon atom, blocking the flow of electricity down another chain. In a sense, the carbon atom functions as a valve, shifting the "open" channel one position with each soliton.

Electrical engineer Michael P. Groves of the University of Adelaide has drawn up a schematic diagram using these soliton valves to simulate information gates for the Boolean logic variables AND, OR and NOR.

However, actually building anything using Carter's ideas is far beyond the capacity of current technology and will probably present numerous technical and financial problems for many years to come. As University of Pennsylvania biochemist Joseph Higgins says: "Carter has figured out all sorts of things you could do with soliton systems, but it's all paper chemistry. He first has to demonstrate that they really exist."

Even if the underlying theory of solitons that powers Carter's visions proves true, the notion of microelectronic switches based upon single conductive molecules faces vast developmental impediments. The most looming is the central question of how it would be possible to interface circuits of single molecule size to other systems. Current computer circuits use fine wire leads to connect microcircuits with the macro world around them. But no wire would work with a single molecule. Not even laser light would work, because the wavelength of the beam would be wider than the space between densely packed molecular elements. And expanding the spaces between elements would defeat the primary purpose behind biochip development. No one knows exactly how to address this question.

Single molecule switches would also be subject to random electron tunneling among closely packed elements. This would create a "short-circuit" situation, destroying data and the circuit's usefulness. Again, spacing out elements to prevent this possibility would limit the miniaturization possible with these switches.

And, single molecule switches would probably be fairly unreliable. At these small sizes a certain percentage of transcription errors and huge effects from infinitesimal physical changes — such as heat, low-level radiation or cosmic rays — seem inevitable. Even today's silicon chips are subject to occasional soft errors, making quality control a significant headache for fabricators. This problem could be even worse at the minute specifications of molecular switches.

To address some of these problems, a few researchers have suggested leaving single molecule technology in favor of semicrystalline switches containing thousands of molecules that can shift in concert.

Richard Potember of the Johns Hopkins University Applied

Physics Lab, for instance, has worked with a thin film of polycrystalline TCNQ in a complex with copper atoms. A sandwich of copper, TCNQ and aluminum, the TCNQ-copper crystalline combine splits, in the presence of high voltage, into its component parts, owing to the shift of an electron. This results in the material moving from high to low resistance. The process takes less than five nanoseconds and represents a binary state that could provide a crystalline switch. High-energy laser illumination has a similar effect, so these switches might even be driven by light.

Potember's most recent work has resulted in the creation of an optical switching system that utilizes organic molecules for the switch and laser beams as the means to flip them. He also is looking into organically based, incredibly high-density erasable optical storage disks.

A Naval Research Laboratory team has obtained effects similar to Potember's by exposing copper phthalocyanine molecules to circularly polarized yellow light. While crystal switches, by being redundant, get around reliability problems, and by being larger, obviate the interface dilemma, they also wipe away the principal benefit for biochips, as well. They don't achieve the vastly increased densities that spurred molecular switch thinking to start with. In fact, they could approach the size and performance characteristics of today's microchips.

Many researchers feel that even if molecular digital switches were possible, they wouldn't represent the best use for molecular technology in computer circuits. These scientists feel that the coming requirements for robotic spatial comprehension, artificial intelligence and concurrent processing fit much more neatly with analog biocircuits.

During the next few years, computers, in order to handle the coming level of complexity in the problems they will face, will have to break through the so-called von Neumann barrier. Mathematician John von Neumann postulated, in the early years of computer development, that computer circuits had to handle operations one at a time, sequentially, through a single computational channel. All computers today have conformed to that model. But the rigidity of von Neumann machines makes them unsuitable for artificial intelligence applications or for the massive processing that some future computer problems will require.

Analog computers don't suffer from the von Neumann lockstep, but they have been neglected in computer development because of their size and lack of programmability. Looking ahead, however, theorists such as Tim Poston of UCLA's Crump Institute for Medical Engineering, perceive molecular analog computers that could use the shapes and contours of long incremental information demanded by AI, robotics and parallel processing. He says: "To interpret a natural language sentence, or to translate a metaphor appropriately, is to model human behavior. This requires an isomorphism, at some logical level between brain function and computer function. We cannot redesign the brain; we may have to redesign the computer."

The new analog computers would use long, chainlike molecules, such as proteins or cell-surface receptors for hormones, and would store information through the molecule's shape rather than electrically or electrochemically. This ability to store in 3-D makes these analog switches "more intelligent than simple on-off devices, capable of reacting with greater nuance and flexibility. Using control chemicals, pH and other factors, these molecules could be switched among several different states, rather than just two." F. Eugene Yates of the Crump Institute says: "Enzymes are capable of saying not only yes and no, but maybe."

A drawback of analog molecular switches is that they would perform rather slowly by computer chip standards, and that the processes they performed would have to be limited; they would still be subject to the nonprogrammability limitations of analog systems. So they probably couldn't replace digital circuits for basic computer operations. But for the motor control requirements of robotics or the inferential demands of an artificial intelligence computer they could provide superlative partners for the basic operational switches.

An example of the role an analog biochip might take would be in the situation of a robot capable of hitting a baseball. Tracking the trajectory of the ball could easily be accomplished with a traditional algorithmic chip. It could "remember" the size and location of the strike zone and issue a swing/no swing decision.

But it couldn't deliver the amount and inflection of information required to actually move the bat through the ball and produce a clean hit. The multiple positions of hand, arm, body, bat, ball, the shifting of weight is too dynamic a process to be handled in rigid, lockstep fashion. The analog chip could better handle this task, making the slugging robot an analog/digital blend.

Using robotic vision as a case study, Michael Conrad of Wayne State University has postulated a working arrangement for translating data into analog biochip form. His system begins with a TV camera image that is digitized and converted into individual tiny dots, known as picture elements or pixels. This dot pattern would interact with a layer of photosensitive chemicals that would produce a gradient chemical reaction based upon the intensities and color of each dot.

Next, this gradient chemical flow would pass over an array of protein based switches secured in an underlying membrane. The proteins would be activated in a gradient pattern analogous to that of the chemical. Beneath the membrane, the protein switches could stimulate a single enzymatic reaction or a "cascade" of them, building up and refining the pattern of reaction.

Finally, minute biosensors would read the enhanced bio-chemical message and would generate analog electrical waves that would power and control robotic actuators, vision mechanisms or other systems.

In theory, these processes could lead to the massively parallel, graded information-handling pattern of the human brain, resulting in smoother motion, greater pattern recognition and many other benefits for high-level computing and robotics. But critics assert these concepts are vague, and the practical underpinnings of them are, as yet, virtually nonexistent. Still, Conrad asserts that the flexibility inherent in analog chips makes it worth the struggle to actualize them, because they could provide the medium for creation of truly intelligent machines that could organize the universe in a human-recognizable form, adapt and internalize the patterns of past behavior.

On a slightly more prosaic level, protein molecules have intrigued other researchers, not because of their shape, but for more fundamental reasons — the long chains of genetic codes they contain. To scientists such as Akiyoshi Wada of Tokyo University, the four components of the genetic chain provide the possibility for complex molecular "software" that is synthetically produced and carries messages that can be retrieved and used by computer circuits, but is also alive and nontoxic for human systems.

"Protein," says Kevin Ulmer of Genex, a biotechnology company, "is the basis for everything from bacteria to whales — the diversity is enough to say that we can do an awful lot of intersting things with it. We should treat protein as a novel material and develop techniques for manipulating this material."

"The computer revolution is going to continue for generations," says Dr. James McAlear, director of the Center for Advanced Research in Molecular Electronics at Texas A & M University, and one of the most zealous proponents of protein-based switches in the biochip pantheon. With the help of proteins, McAlear says, "the difference in speed and density between today's integrated circuits and the molecular chips of the future will be greater than the difference between the vacuum tube and the integrated circuit."

McAlear and his development firm, Gentronix, are working with Texas A & M, trying to find applications for molecular electronics in industry. Once recent prototype being tested had circuits etched in a layer of the protein polylysine, rather than in the purified silicon used today. In tandem with John Wehrung, a former Sperry Corporation engineer, now president of Gentronix, the company holds what McAlear calls a "Wright Brothers" patent on the new technique, which he likes to compare with the first flight at Kitty Hawk. In McAlear's words: "Our method combines two technologies — biophysics and electrical engineering — in a very simple way, like adding an engine to a glider."

The idea underlying the process is simple enough: McAlear and Wehrung use a film of synthetic protein covered with a layer of plastic. They carefully strip off the plastic cover in the pattern of a microcircuit. The exposed protein is then washed with a solution of silver nitrate, which it precipitates. By design, the protein then reduces the common photographic chemical to silver. The result is what Wehrung calls an "electroless plating operation," depositing metal wires by an organic chemical reaction. Traditional silicon chips require volatile and often dangerous inorganic chemicals to get the infinitesimal wires in place.

Though the density of circuits resulting from the Gentronix technique is not much greater than conventional silicon techniques, Wehrung says that "the critical advantage lies in the ability to deposit novel metals one layer on top of another." Another chemical bath provides a "bridge" between the metal layers. By creating three-dimensional structures with multiple points of contact between the layers, McAlear and Wehrung believe they have fashioned an inexpensive material base for parallel processors, the kind of incredibly high-speed chips that will be required by the next generation of computers.

McAlear and Wehrung admit, though, that the future of computer chips goes far beyond their unique technique of depositing metal. They believe that light is yet another untapped source of microchip circuitry. In fact, they hold a second patent on a means of introducing a crystalline pattern to molecules in a protein layer whose conductivity changes in response to light.

These research ideas promise drastic reduction in the size of computer components. To dramatize the miniature proportions of the new media, McAlear compares his own protein-based circuit to a boy three feet tall; in the same scale, Dr. Carter's molecular computer memory would be comparable to the thickness of the boy's fingernail. And today's microprocessor would be a gross hyperthyroid case.

While McAlear's switches are exciting, some of his other projects and speculations raise possibilities for biochips that go far beyond any other scientist's conceptions. McAlear sees the possibility of actually interfacing protein-based computer circuits and the electrochemical circuits of the human brain.

The first expression of these ideas came in McAlear's 1981 National Science Foundation project to design an electronic implant to bring sight to the blind. Metal-based electrodes have been used in the past to give a vague sense of sight to sightless people, but metal wires can link only with clumps of nerves, not with single strands, and the implanted metal is an ever-present danger for the bearer. McAlear plans to build an array of 10,000 minuscule electrodes, encasing them in protein and plastic film. Using an electron beam, he will drill infinitesimal holes through the plastic to the protein above the electrodes, then dip the device in a culture of embryonic nerve cells. When implanted, McAlear believes, the embryonic cells will connect with single neurons in the visual cortex, thus linking them to the electrodes. An external TV camera would send images to the implant, which could then send signals down the connected nerves to produce a sightlike image in the person's brain. "We hope to have the device ready before the decade is out," McAlear states.

A similar system is being developed by neurobiologist Jacob S. Hanker of the University of North Carolina. His researchers have managed to deposit live cells onto a silicon chip. "If specialized cells, such as nerve cells, were attached to a chip," he states, "they could perform specialized functions, such as detection and monitoring."

Beyond these intriguing possibilities, McAlear speculates about the creation of an entire molecular computer that could be implanted in the human brain. He points out that programmable insulin pumps and other drug delivery systems are already in use. His implantable computer could serve as a central control center for people with chronic illnesses, or as a formidable diagnostic tool. It might even have possibilities for specialized learning.

And, McAlear also wonders about what happens when entirely biologically based circuits — especially those based upon the DNA code — reach a point where they can be programmed to replicate themselves. "We then come to the question of whether or not this thing is alive. This would be intelligence evolving not through natural selection, as man evolved, but through research and development, using its own computing capacity to improve itself. Of course, this is way, way in the future." For now, McAlear and other biochip scientists are careful not to oversell the immediate benefits of their work. Generally, they predict that biochips and molecular chips will have their biggest short-term impact at the high ends of computing, where chip density is becoming a crucial concern. But they feel the benefits of biochips will move into homes quickly, because of their potential for bringing computer capabilities into any nook and cranny — an appliance, a light, a bicycle.

When trying to estimate the long-term pace of biochip and molecular development, scientists again are cautious about making any promises. They point out, for instance, that no fully practical system for fabricating any of the biochip systems currently exists. Proposals range from jigsaw puzzle plans for building up switches in pieces within specially designed cells to utilizing ordered molecular films that lead protein subunits to assemble in predictable spatial patterns. McAlear has had thoughts about using the unique properties of monoclonal antibodies to create corresponding "switch" proteins that can have only the single orientation accepted by the monoclonal. Still another idea is to use crystalline structure as lithographic-like masks, a single atom wide, to lay the paths for the molecular wires to follow.

All of these concepts, however, lie on the edge of scientific possibility. Still, McAlear points out, the rate of technological advance has been exceptionally rapid in the past two centuries. "In every major technology, like aviation or space flight," he says, "the wildest optimist has turned out to be far too pessimistic."

"I tend to be a pessimist, so I think of twenty, forty, maybe fifty years. But there may be shortcuts that bring the molecular computer into perspective within the next decade." In the words of computer industry seer John Diebold, who first coined the word automation, what we are approaching is "the inevitability of the convergence of biology and . . . information technology, first through applications of the insights gained in neurobiological research to the organization or architecture of computer-based systems and eventually the possibility of biological computers formed by programming the stuff of life — DNA."

Chapter Five

Communications:
Around the World in Split Seconds

By Andrew Pollack

Andrew Pollack is a technology writer for The New York Times.

In a remote part of New York's Staten Island, a new port is in operation. No ships ever visit it, though, for it is not even near the water. Nor do goods or people move into and out of New York City through it, because this is a different kind of port — a teleport, a port of information in and out of a great metropolitan area. The information (telephone conversations, computer data, television pictures and the like) is shuttled between New York office buildings and the teleport by hair-thin strands of glass. Once at the teleport, the data are skyward to satellites, to be whizzed around the country and the world. The teleport was built in part by the Port Authority of New York and New Jersey, an organization normally concerned with bridges and tunnels, marine terminals and airports. That such an organization took on a project like the teleport is a sign of a fundamental shift occurring among the peoples of the world.

The advanced economies are moving into the postindustrial

era, one in which the movement and production of information are becoming as important as the movement and production of goods. Already, more than half the workers in advanced economies like the United States and Japan spend most of their time gathering, manipulating, analyzing and distributing information, rather than producing goods or services.

But what does it mean, that information flow becomes as important as the movement of goods and people?

It means, for instance, that people may work or bank or shop entirely from their homes, using computers connected through the telecommunications channel to other computers.

It means that executives can have telemeetings, using video and telephone systems, instead of traveling to business meetings; that engineers in Europe and those in America can cooperate on designing a new product using graphic design information relayed from computer to computer.

It means that, rather than manufacturing products in one central location and then shipping them out, the information governing production might be sent out and the item produced where it is needed.

It means that purchases can be made by transferring electronic blips from the buyer's account to the seller's account, without any cash or paper changing hands.

So far, however, many of these innovations have taken longer than expected to catch on. People are not willing to give up face-to-face meetings for a video screen, nor are they willing to work at home or shop via computer. The age of information has a fair degree of hype to it. Nevertheless, it is undeniable that the movement of information is becoming extremely important. And that means that telecommunications networks are becoming as important in the information age as railroads and roads are in the industrial age. Telecommunications networks are the superhighways of the next century.

To meet the challenges of the new age, the telecommunications technology is undergoing a massive transformation. The news has been filled with reports of developing technologies — car telephones, television from satellites, optical fibers and high school computer whizzes with home computers electronically rummaging through a big corporate computer on the other side of the country. All are examples of the growing sophistication and scope of telecommunications. On top of the technological changes are institutional changes. The United States, Britain, Japan and Canada are all taking steps to deregulate telecommunications and bring about a new era of competition and innovation. As Charles L. Brown, chairman of American Telephone and Telegraph Company, said on the eve of the breakup of his company: "Today signals the beginning of the end of an institution — the 107-year-old Bell system — and the start of a new era in telecommunications in this nation."

THE NEW ERA

There are so many new developments in telecommunications that it is difficult to sort them all out. Nevertheless, certain basic trends are emerging and override all the developments.

Perhaps it is best to start with the situation that existed around 1970, before many of the rapid changes began. At that time, there were separate systems for the various forms of communication. Voice, of course, was transmitted through the telephone and radio. Printed data and words were transmitted through a separate system, generally through telex machines. Video was transmitted by television broadcasts. Images were transmitted through the telephone line by facsimile machines. Computers, which churned out data, hardly communicated at all, except to print out their results or display them on a nearby terminal screen.

Another characteristic of telecommunications at that time was that the transmission systems were like empty pipelines. Information emerged from the pipeline the same way it went in, except for some loss of quality because of static on the line. But the information was not modified in any way, delayed or rerouted.

The new, emerging telecommunication systems are quite different. Different types of information are being combined and intermingled on the same pipeline. Telephones, computers and television — once completely separate industries — are merging. One takes on the function of the other. A videophone conversation is like television, in that it involves video, yet like telephone, in that it goes between two people rather than being broadcast. Data can be sent by television broadcasts or by the telephone. Computer-telephone combinations are appearing on the market that allow two people to display data and words on their screens and discuss them over the phone at the same time. Japan's Nippon Telegraph and Telephone Corporation developed a device called the Sketchphone, that allows two people to draw images for each other while talking on the phone. The image drawn at one end appears on the screen attached to the phone at the other. All these developments promise to make telecommunications far more versatile and useful.

The second major trend is that, instead of merely transmitting information unaltered, telecommunications systems are processing the information as well. Information can enter in one form and emerge in altered form at the other end. A computer screen of information can come out as a paper document on a facsimile machine at the other end. Eventually, it might be possible for someone to speak English at one end and have the words come out in Japanese at the other end. Information can also be stored in the network and transmitted at a later time, just as information can be stored in a computer. In short, computerized functions are being applied to telecommunications systems. Indeed, it is almost impossible to tell where the computer ends and the telecommunications network begins.

These developments are made possible by new technology — digital transmission. Currently, most information is transmitted in analog form, as a series of electrical waves. But in the future, more and more information will be transmitted in digital form, as a series of ones and zeros, the same way computers process information. Transmitting information as bits allows the information to be processed in transit, just as computer data are processed. Moreover, it matters not to the telephone wire or cable or radio whether the bits stand for voice conversations or video or data. Thus, all different forms of traffic can be easily intermingled on the same line.

The combination of telephone and computer and television technology is more powerful than any of them individually and promises to transform society. The French have even coined a new term for it: *télématique.*

Ultimately, telecommunications engineers envision broad digital highways providing all types of information transfer. This is the integrated services digital network. Japan's NTT is hard at work developing such a system, which it calls the information network system. Already, the backbone of the system has been constructed — a fiber-optic transmission system running virtually the entire length of Japan from north to south. Eventually, probably not until the twenty-first century, such transmission capability is supposed to be extended to every home and office in Japan, bringing a cornucopia of information services. Some Japanese proponents of the efforts have predicted the coming of a "teletopia."

In Europe and the United States as well, trials of the integrated services digital network are underway. Late in 1986, the McDonalds Corporation began wiring up its offices in the Chicago area to transmit voice along with data and images. Pacific Bell Telephone Company began a test of its system in a northern California community. While homeowners did not have much use for transmitting video, one benefit of the system was that a single telephone line could handle two calls at once, with some channels left over for computer data.

The integrated digital network, however, is years away from true realization. Until then, there will be advances across a broad front. This chapter will survey some of them. Virtually everything that will be offered in the telecommunications of the early twenty-first century is available in some prototype form today. The key questions now are not what can be done, technologically, but what can be done economically and in a way that will be accepted by the public.

THE SMART TELEPHONE NETWORK

The concept of combining the telephone and computers is perhaps best illustrated with the example of conventional telephone service. Normally, when someone places a call, the phone system merely makes the connection. Either the call is completed, or the phone is busy, or there is no answer. To talk, both people must be present at the same time. That is the old system.

Now take the "smart" phone system, with computerized intelligence. The intelligence could rest in the phone itself or in the network. The new central office switches, big machines that route telephone calls through the network, are becoming digital. They are essentially large computers. The same features are also becoming available within large corporations that have their own internal switchboards, known as private branch exchanges.

The smart phone network does not merely try to make a connection. It has a storehouse of knowledge telling it where to route calls. If the line is busy, the system allows the customer to put one call on hold to answer another, a feature known as call waiting. If the call is unanswered, the system can look in its computer memory for a new number to forward the call to, a feature known as call forwarding. Both those features are already available in areas served by digital central office switches. In the future it will be possible for a customer to program his whole daily schedule into the phone network so that calls will automatically be forwarded to him at the proper place. Yet another feature in the future might allow companies with nationwide outlets, such as a fast-food chain, to have a single nationwide number. Whenever somebody dialed that number, the phone system would connect the person to the nearest outlet that was open at the time.

Some customers of Bell of Pennsylvania are getting a different taste of the future. They are testing an experimental service known as CLASS calling. This service, for instance, allows them to know who is calling before they answer the phone by displaying the calling number on a small display screen. How nice it would be to know in advance that the caller was a bill collector so one might not answer the phone! The service allows the customer to program the system to block out certain calls altogether, and to give special rings for other, desirable calls. Or, one can have the phone dial back the last number that called.

Still other features utilize computer technology to make dialing easier. A single-digit or two-digit code can be dialed, and the computer can dial the rest. Or, the phone system can be programmed to keep trying a number periodically when the number dialed is busy or not answered. These features are already appearing in telephones themselves, as well as in the network in Bell of Pennsylvania's test.

VOICE MAIL

We have also said that the system of the future will be able to process information as well as transmit it. By converting voice into digital bits, the voice can be manipulated just as data can. For instance, voice can be easily encrypted, to fight wiretapping.

More important, voice can be stored by computer, just as data are. These systems, now being used mainly by large companies, are known as voice store and forward systems.

One function they serve is similar to that of an answering machine. But they also allow messages to be stored for transmission at a later time. A user of such systems might speak the message into the phone: "Bob, don't forget to pick me up at the airport in an hour." He might then fly to his destination. At one hour before his arrival time, the voice store and forward system would dial Bob's number and transmit the previously recorded message. This is, in effect, a voice telegram.

Another use is for mass voice mailings. A manager, for instance, might dictate a single message and have the system dial all of his salesmen around the country and transmit the message. Managers can store copies of the message in their own voice mailboxes. Voice mail systems are thus, as their name implies, a hybrid. They use voice as the means of conveying information, but do it in a matter similar to message delivery. The sender and the receiver no longer have to be present at the same time.

One drawback of voice message systems now is that they are expensive. Storing one second of speech with the highest-quality digital technique requires the same amount of computer memory as more than 1,000 printed words. However, computer memory costs are constantly falling. In addition, data compression techniques are under development aimed at reducing the number of bits needed to store speech in digital form.

The ability to store voice might lead to new forms of communicating. At Bolt, Beranek & Newman, a Cambridge,

Massachusetts, consulting firm, researchers are experimenting with what they call compound documents or multimedia documents. These are not paper documents at all, but pages of a computer screen. The pages contain text and diagrams, like paper documents. But the documents can also contain computer programs and voice messages. By moving the cursor to a particular symbol, the voice message is activated. These compound documents allow someone to send what is essentially a slide show through the telephone system. The only way now to send a slide show is to mail a tape recording along with the slides.

In the future corporation, documents might be passed electronically from worker to worker for comments. Instead of scribbling their comments in the margins or on a separate piece of paper, as is done now, the workers might be able to make their comments by speaking. The voice comments would be just as much a part of the document as the written comments. The voice would be filed electronically with the written comments and transmitted electronically. Whenever somebody retrieved the document from storage and put the cursor on the proper symbol, the voice comment would be heard. This is one example of the overriding trend that different forms of communication are merging.

We have seen, then, that applying computer technology to the telephone system promises greatly to increase the versatility of the system. It frees two people from having to be at their phones at the same time to communicate. Perhaps even more significant, however, is what can be done by applying telecommunications technology to computers.

DATA COMMUNICATIONS

When computers were first invented, they were large and complex beasts confined to special rooms. Now however, computers can be small and portable, and computer power can be embedded into other devices, from microwave ovens to missiles. These smaller computers can communicate with each other over the telephone lines or other lines. It is the combination of telecommunications and computers that has the potential for transforming society, more so than each of the two separately.

So far, communications among computers is not as fluid as it could be. In the simplest means, a small computer or terminal merely connects with a larger one through a direct telephone connection or through a special dedicated connection. The little computer can then send information to the larger computer or retrieve information from the central computer. One problem is, however, that telephone lines are designed to transmit tones, not digital bits. So devices known as modems must be used at either end. These devices convert the computer pulses into analog tones that can be carried by the phone system. But these add expense and slow transmission speed. The usual speed of communications is either 300 bits per second, or about 30 alphabetical or numerical characters a second, or 1,200 bits a second, equivalent to 120 characters a second. Faster speeds, up to 9,600 bits per second, are used with special phone lines.

Another difficulty of connecting computers over the phone line is the expense involved. A long distance telephone connection is wasteful for computer communications. A person using a computer to retrieve some information from a large central computer, for instance, might transmit a request for certain information. The bits or data representing the request travel down the line in a fraction of a second, and the information requested is sent back from the computer, also in a few seconds. The person might then spend several seconds or even several minutes reading the information before making another request. All that time, the phone line is not being used, although the user is being charged for it.

A solution to that, which has already lowered the cost of computer communications significantly, is packet switching. With packet switching, no circuit is set up for the exclusive use of two participants, be they two people or two computers. Rather, messages are broken up into little packages, called packets. Each packet is given an address, like the address on an envelope, and sent through the network by whatever circuit is open. The packets from many different computers are intermingled on the lines, just as cars traveling from Detroit to New York might intermingle on the same highway and stop at the same gas stations as cars traveling from Chicago to Philadelphia. This is a lot cheaper than building a special road from Chicago to Philadelphia that will have only an occasional car on it. The two major packet-switching networks in the United States are Telenet and Tymnet, although there are many others.

Yet another development in computer communications is the local area network, or LAN. This is a system designed to connect the many small computers within an office or a building or a small complex of offices. With a local network, little computers no longer need to have their own expensive printers and disk data storage systems. They can share common ones and draw on common data. In addition, the little computers can be used for sending memos within the office electronically. Local area networks are usually built with a "gateway" connecting the local network to long-distance networks, such as the nationwide packet-switching systems.

While both local networks and packet switching have made computer communications more practical, there are still numerous obstacles to achieving fluid communications. The problem is standards. Different computers made by different manufacturers cannot understand one another. Network standards provide for rudimentary connections. It is possible, for instance, to send a simple text message from one computer to another using a common bit code, called ASCII, to represent the different letters. But the computer command to italicize a word on one computer might not be understood by the other computer. Graphic images made on one might not be easily sent. It is analogous to the phone system. The world telephone system is standardized to the extent that one phone anywhere in the world can connect with another. But if the people at either end speak different languages, they still cannot communicate.

Efforts are under way now to develop so-called higher-level standards for computer communications, meaning those standards beyond the basic connection. Progress has been slow because it has been difficult to get companies to band together. The General Motors Corporation, one of the world's largest users of computers, was so frustrated that it designed its own standard, known as the manufacturing automation protocol, or MAP, to connect computers, robots, electronic testing equipment and other devices within a factory. Most computer companies, fearful of losing the automaker's business, say they will go along with MAP. Computer makers have now also banded together to form a group known as the Corporation for Open Systems to develop standards for office communications. But the effort is likely to take years.

Despite the fact that computer communications still could stand a great deal of improvement, even the existing capabilities are extremely powerful. The factory, office and home are all likely to be fundamentally transformed by computer communications. The transformation is occurring already, as the following examples indicate.

ELECTRONIC MAIL

With one computer connected to another, one obvious thing that can be done now is to send messages from one screen to another. This is far faster than using the mail. If the person is not at the other end to receive the message as it comes in, the message can be stored in an electronic mailbox. Electronic mail poses a long-term threat to conventional mail. However, right now such systems are not catching on because most people don't have computers.

DATABANKS AND ELECTRONIC PUBLISHING

A vast wealth of information is now stored in central computer depositories and can be retrieved by terminals or small computers. Using key words, it is possible to extract from the vast amount of data only those items of particular interest. There are now more than 1,000 on-line data bases in the United States. They include everything from chemical abstracts, to legal cases, to information on horse breeding and lists of stolen books.

While many are expensive, costing several dollars per minute to use, consumer-priced data bases are coming into existence, spurred by the spread of personal computers. Already on-line are an encyclopedia, news reports, airplane schedules, classified ads and restaurant reviews. In France, consumers are being given terminals instead of telephone books to retrieve listings from a central computer.

Sometimes, these electronic information services are called videotex. Videotex usually refers to a particular method of transmission using frames of information sent to specially adapted television sets serving as terminals. Videotex has not caught on yet, and some initial attempts at it, such as the Knight-Ridder Company's Viewtron service in Florida, were disbanded when a market failed to materialize. These disappointing early results seem to contradict the notion that newspapers and books will one day be replaced by computer screens. Books and newspapers are more comfortable to read and easily portable. Computers now cannot easily reproduce photographic quality pictures. It takes a long time, at current transmission rates, to transmit a photograph electronically.

But computers might take over part of what is provided by newspapers. This is especially true where the computer offers advantages that paper publishing doesn't. For instance, it is wasteful to print the entire stock table when each reader is usually only interested in a few stocks. It might be more efficient for the individual to obtain stock prices from a data bank but his general news from the newspaper. Such systems become even more attractive when the full power of the computer is used. What if instead of merely retrieving stock prices, the computer automatically inserts the new prices into a model of the portfolio and calculates the new value of the portfolio? The home computer might also flash a warning that a particular stock has risen above a certain level, meaning it is time to sell. In that case, it makes far more sense automatically to retrieve the information by computer, rather than from the newspaper. Indeed, many investors are already doing this. The choices, however, are not merely between traditional print publishing and computerized data banks. There are several hybrids that are also catching on.

One is to distribute data banks by traditional methods, like the mail. A computer disk containing the data bank can be mailed out to users. This saves the user from having to connect over the telephone line each time information is wanted. Standard &

Poor's, for instance, distributes its data base of stock information on computer disks. A particularly useful way to such distribution is optical disks. The same compact disks now used for music can be adapted to store 550 million characters of information, enough to store a quarter of a million typewritten pages. Already Honda dealers are using computerized parts manuals, replacing the volumes of binders that would fill a whole shelf. Telephone directories, encyclopedias and huge library card catalogs are also being put on disks. The one drawback of this scheme is that the information is not up-to-the-minute, as it is in a central data base.

The second hybrid is distribution of information via telecommunications so that it is printed out at the other end. Instead of reading a newspaper on the screen, a homeowner might find the newspaper printed out every morning by his home printer. This can also be done for other types of information. Software can be transmitted electronically and stored on a disk, rather than sold through computer stores. Music can be transmitted and stored on tape, something that is already done on an ad hoc basis by many consumers. The only difference is that now the consumers cannot make specific requests for what music they want broadcast or transmitted. All these types of transmission are known in general as downloading. For printing, the term printing-on-demand has been used and is already being experimented with, particularly for technical publications. Rather than print thousands of copies of a book and storing them in a warehouse, some publishers are storing texts electronically. When a customer wants one, the electronic signals are sent to a special printer, and the manuscript is printed out.

Yet a third hybrid is desktop publishing. Here a computer and a laser printer are used to do page layout for traditional print publications. The computers make page layout so easy that numerous little publishers, who once might not have had the money or the wherewithal, are now publishing little newsletters that look almost professional.

COMPUTER CONFERENCING AND BULLETIN BOARDS

When a Korean airliner was shot down by the Soviet Union in

September 1983, the conference rooms at Participation Systems Inc. sprang to life. People began discussing who was responsible for the shooting, whether the plane was on a spy mission and how much the United States government knew about the incident. The interesting thing, however, was that the people holding the discussion were not really in the same room at all. They were spread all over the country and were not even all talking at the same time. Nor did they even know one another.

Participation Systems was offering a computer conferencing service. Computer conferencing is another new type of communication made possible by the combination of computers and telecommunications. A computer conference is like a data bank, except the information is put into the data bank by the users themselves. A participant in a computer conference connects with the central computer and reads the dialogue since he was last on the system, then he adds his comments or answers questions left by other users, and so on. Most computer conferencing systems also allow for a user to communicate in private with another participant, just as if the two participants walked out of a real conference room into the hallway to have a private chat.

The power of computer conferences is that, for the first time, it is possible to have an ongoing dialogue without all the participants being in the same place, or without them even being assembled at the same time, which would be necessary for a telephone conference call or video conference. Such conferences have been used by lawyers from around the country planning a defense and a group of authors cooperating on writing a book. Nuclear power plant operators maintain an extensive conferencing system to discuss problems they encounter and possible solutions. Indeed, virtually any organization in which users share a common interest can benefit from such a conference.

The consumer-level equivalent of a computer conferencing system is the computer bulletin board, which is the electronic equivalent of the supermarket bulletin board. These are home-grown systems, often set up by computer hobbyists using their own personal computers as the central computer. Users call up and leave messages or read messages left by others. There are several thousand such bulletin boards now in use in the United

States, although many of them last only a short time. Clubs are finding them useful for their members to exchange information. There are some bulletin boards on which users seek dates. There are also some so-called pirate bulletin boards on which illegal information is exchanged, such as telephone credit card numbers and code words for breaking into corporate computers.

TELETRANSACTIONS

The same systems that allow for retrieval of information and computer conferencing also allow for transactions to be performed remotely. Instead of merely reading the airline schedules on a computer screen, for instance, one can make reservations. Such teletransaction systems are in various stages of development, but all will have some problems winning full acceptance. Here are some of those that are being tried:

Telebanking. With home computers, consumers can check the amount of money in their accounts, pay bills, and transfer money from one account to another. Some home banking systems allow for automatic monthly payments, such as for utility bills or mortgages. The two things people cannot do with telebanking are to make deposits and withdraw cash. These are very serious shortcomings and are a major reason that home banking so far has not caught on.

Teleshopping. In these systems, consumers can place orders after scanning electronic catalogs. The customers can specify what types of products they are interested in and the price ranges and characteristics. They are then shown descriptions of products that fit the description. Teleshopping offers convenience but has its drawbacks. While it is fine for certain standard items, like televisions, it is much more difficult to shop for soft goods, such as clothing or furniture. People like to touch and feel those before buying. Currently, shop-at-home services offer text-only descriptions. In the future, they can offer photographic-quality images and, perhaps, even brief video segments. As of now, many customers use the teleshopping for comparision shopping, to find out prices. Then they go to a store to make a purchase.

Televoting. Electronic communications can be used for vote taking and poll taking, allowing for virtually instantaneous referenda on public issues. So far, this has been tried only in a limited way, using two-way cable television, not computers.

Teleinvesting. Electronic communications can prove a boon to investors. They can program their computers to contact a data base, retrieve prices of stocks in their portfolio and perform analyses of the trends. Some brokers allow the clients to place orders for buying or selling stock by computers. Still, such on-line broker systems have not proved very popular yet.

The cashless society. Electronic transactions need not be confined to the home. Already, much of the world's commerce involves payment by transfer of electronic blips rather than the actual movement of cash. In the future, this use may spread. Gasoline stations and grocery stores are already beginning to allow customers to pay with bank debit cards — the same ones used in automated teller machines — that automatically transfer money from the customers' accounts to the stores' accounts.

Businesses are also likely to be big users of other electronic transactions. A grocery store's point-of-sale terminals can now automatically tabulate what goods are being sold and what's left on the shelves. In the future, when the system calculates that the store is running low on mushroom soup, the store's computer can automatically contact the soup company's computer and place an order. The payment could also be made electronically.

Such electronic transaction systems are often called electronic data interchange (EDI). Already it is being used by the grocery business, railroads and the auto industry. Electronic transmissions from computer to computer take the place of routine business forms such as invoices and purchase orders. George Klima, director of accounting systems for Super Valu Stores, an early user of EDI, said it costs the food wholesaler and retailer $1.75 to call in a purchase order and send a paper confirmation. Electronically, it can be done for thirty cents.

If EDI spreads, as it will, it will mean that those companies without computers will be as cut off from the flow of commerce as a company without a phone would be today. The Ford Motor Company has already eliminated paper purchase orders in some divisions. General Motors has already told its auto parts suppliers, in effect, that they must be able to receive orders from the auto giant electronically, or they risk being cut off as a supplier. Still, developing EDI requires an industry to agree on standards, which can take a long time.

Another use of telecommunications combined with electronics is monitoring and controlling appliances remotely. For instance, instead of sending a meter reader to each house, an electric company might have meter readings sent remotely. This is telemonitoring. The utility might also be able to send signals, either over telecommunications circuits or through the electric wires themselves, to control the operation of appliances. This is telecontrol. Some utilities, for instance, have already experimented with telecontrol systems that allow them to turn off some customers' hot-water heaters at times of peak electricity usage. By reducing peak demand, utilities can avoid having to build expensive new plants.

There are other uses of telemonitoring and telecontrol. One is security systems. Such systems generally connect with central command posts, or the police station directly, over telephone or two-way cable television. Oil-pumping locations, dams and vending machines can also be controlled remotely.

Such systems can also be used within buildings. Buildings that employ such systems have been dubbed intelligent buildings. Sensors placed in rooms can indicate when rooms get too hot or too cold, when intruders have entered, when lights are left on at night and when there are breakdowns in water or electrical systems. In the Hilton hotel at Disney World Village Plaza in Orlando, Florida, guests control room temperature and even the television set, through the telephone.

The next step after the intelligent office building might be the smart house. In this, all appliances would be controlled from a central panel, or even from outside the house by dialing the control panel. There are already devices that allow this to be done. One can call up one's home and start heating up dinner before one arrives, or program the air conditioner to turn on or off. Still, such systems are not likely to catch on for a while. One problem, as always, is developing standards for home appliances to be hooked up to communications network. Yet another problem is that there seems to be little advantage to central control of home appliances.

So far, the changes in communications described have mainly concerned combinations of computers and telephones, with the signals traveling over the conventional telephone network. But equally significant changes are occurring in the means of transmission.

OPTICAL FIBERS —
TRANSMISSION AT THE SPEED OF LIGHT

The most significant of these is the change in transmission technology from electricity to light. Rather than sending electric signals over copper wires or microwave radio links, users will transmit more and more information as pulses of light traveling over hair-thin fibers of ultra-pure glass. The glass used is so pure that if a window pane could be made from the material, it could be twenty kilometers thick and still be transparent.

Because light signals are at extremely high frequencies, light waves can carry hundreds of times more information than copper wires or radio waves. A cable with 144 glass fibers, the thickness of a thumb, can carry at least five times as many calls as a cable of 1,500 copper wires, the thickness of an arm. The contents of an entire encyclopedia can be transmitted in a matter of seconds.

Optical fiber transmission offers other advantages as well. The transmission is free from electromagnetic interference. And signals cannot easily be intercepted by wiretappers. But the main advantage is the huge quantities of information that can be carried. A single optical fiber coming into a home could carry telephone conversations, dozens of television channels, computer data and electronic mail.

Optical fiber technology was developed in the 1960s and early 1970s, but already the transformation is under way. So far, optical fibers are used mainly for high-capacity telecommunications links between computers or between telephone switching centers. By the mid-1980s, extensive networks of optical cables were being laid across the United States, Japan and Europe, and plans were being made to lay cables beneath the Atlantic and Pacific oceans.

The final step would be to bring optical fibers to individual homes and offices. This is likely to take a very long time because it is expensive to replace the existing copper wire, and a lot more information can be sent over existing wires than is now being done. Nevertheless, it is part of Japan's plans for the information network system to do this. Bringing optical fibers to the individual consumers would lay the groundwork for a cornucopia of telecommunications services. In addition to telephone calls and high-speed data, such wires would be able to carry numerous television channels and to have two-way television. People would be able to see each other when they talked on the telephone, or a parent could monitor children playing next door.

Optical fibers also offer promise for use in industrial settings. This takes advantage of the fact that optical fibers can be used in a variety of sensors, as well as for communications. Optical fibers could be used to relay information around a factory floor. Fiber optic sensors could peer into furnaces, where cameras cannot go. Optical sensors could monitor pressure and inspect products coming off the assembly line. In Japan, the Tokyo Electric Power Company is stringing fiber along its power lines to monitor circuit failures and relay the information to control centers. The fiber is ideal for this because light wave communication is immune to interference from the power lines. Nippon Kokan's Fukuama works, the largest iron mill in the world, uses fiber optics to relay information from the oxygen furnaces and blast furnaces to control computers. Hitachi and Toshiba have fiber-optic networks connecting computers.

One problem with optical fibers now is that when the signal has to be amplified, or routed through a switch, the light signal must first be converted into electricity. Then it is switched or amplified electronically and converted back into light pulses for the next leg of the journey.

Scientists are trying to work around that by developing the optical equivalents of integrated circuits. The Japanese Optoelectronic Project is working on an optoelectronic integrated circuit. This chip would combine the light receiver, electronic switching circuit and light emitter on a single chip, lowering the cost of the conversion process. Such a chip would be made not of silicon but of gallium arsenide, a material that can act as both a light emitter and detector as well as a semiconductor. But gallium arsenide is difficult to work with and gallium arsenide chip development lags way behind silicon chip technology.

Others favor making chips that process light itself, avoiding the need to convert the light to electricity. This would require the development of all-optical chips, an even more difficult task. Scientists have barely achieved the creation of an all-optical transistor, a device that can either block light or let it pass through, depending on the presence or absence of a second light. The silicon transistor was invented in 1947, so in a sense optical chip development is only where electronic chip development was forty years ago. If all-optical circuits are developed, however, it might pave the way for all-optical computers that could be extremely fast, yet low in power consumption. Such a development, however, is decades away.

RADIO TRANSMISSION:
GO-ANYWHERE COMMUNICATIONS

One service that optical fibers cannot provide is portable communications. Yet communication by radio waves is already changing the face of communications. A whole range of portable communications devices are being developed. On the positive side, this means that no busy person need ever be out of touch riding on a plane or a train, driving in a car or walking through the streets. The bad side of this, however, is that busy people might never be able to escape from those wanting to reach them.

The biggest new development is cellular mobile telephones, which began spreading through the United States in 1984. This technology promises to allow for a vast expansion in mobile telephone service, which heretofore has been restricted to a few customers because the available frequencies were overloaded. In New York City, for instance, only 12 mobile phone conversations could take place at the same time.

Cellular technology can change that and allow for virtually unlimited mobile phoning. Instead of one central transmitting and receiving antenna serving the whole city, the region is broken down into hexagonal cells, each with its own antenna. Thus the same twelve New York City channels that once served the whole city can be used by many different vehicles at the same time, each within a different cell. As more channels are needed, the cells are made smaller. As a car passes from one cell to the next, the call is handed off from one antenna to the next by sophisticated computerized switches. The caller doesn't even notice the change.

So far, cellular phones are expensive. It can cost more than $2,000 for the phone itself and more than $100 a month for the service. Eventually, however, prices are expected to drop. When that happens, car phones might spread to consumers, and a new problem might arise — safety. Talking on the phone while driving, especially if one hand is needed to dial and hold the phone, could lead to increased accidents. However, mobile phone and automobile designers are already designing phones with automatic dialers to dial frequently called numbers when only one button is pushed, or even recognize voice commands. In addition, some automakers envision putting the phone mouthpiece in the visor and using the car's speaker system as an earphone. Then the driver would not have to hold a phone handpiece while talking.

But car telephones are just the first use of cellular technology. It will also lead to portable phones that can be carried around. This could bring about a major change in the concept of phone numbers. Right now, phone numbers are associated with locations. To call someone, one must know where the person is. But with portable phones, the phone number could be associated with a person. Each person would have only one phone number, associated with a portable phone. The person could be reached anywhere at anytime with that number. The phone number would be almost a personal identification number. With continued advances, phones might get down to the size of wristwatches, as in old comic strips.

But portable telecommunications, like other forms, need no longer be confined only to voice. Combined with portable computers or terminals, data and images can also be transmitted. In Japan, miniature facsimile machines have been developed that are placed in police cars. If a suspect is wanted by the police, his image can be instantly transmitted to all patrolling cars.

IBM field servicemen now use portable radio terminals that permit them to receive messages from headquarters, telling them where to go next and allowing them to file reports on their activities and customer problems instantly.

Automobile companies are now testing cars with video screens. One purpose of the screens would be to display fuel, temperature and other readings that are now kept on a myriad of separate dials and gauges. The same screens could be used for retrieval of information, such as traffic reports sent via radio waves.

For those who cannot afford cellular phones and portable communications terminals, the answer might be a pager. The difference between pagers and mobile phones is that the paging signal goes only one way. The person with the pager can receive messages but not send them. Pagers are no longer merely the cigarette pack-sized devices worn by doctors and repairmen, which emit a beep when they are needed. Pagers can do a lot more than beep. Some can deliver a voice message. Others can deliver a written message that appears in a calculatorlike display. Some can even print a message. Moreover, the price is dropping to the point where the use of pagers is spreading to a wider audience. One paging company, in an attempt to develop a broader market, has a "Stork Alert" program in which it rents pagers at introductory rates to families with expectant mothers. The woman can use the pager to contact her husband when it is time for the baby to be delivered or if another problem arises.

Finally, paging heretofore has been confined to one city. But the Federal Communications Commission has allocated frequencies for nationwide paging systems. The signal for the desired pager would be sent to all cities by satellite and rebroadcast locally.

Similar in concept to pagers are special purpose data terminals that can receive information, but not send it. Lotus Development Corporation, a software company, sells a service that broadcasts stock quotes to a terminal or computer. The quotes are broadcast on a rotating cycle, and users program their terminals to "catch" and store the quotes wanted as they come by. This is really a form of radio broadcasting, but it allows the user to specify the information that is desired.

Eventually, as mobile terminals and phones come down in price, they will take the place of pagers. In the meantime, pagers provide a cheaper if somewhat less powerful alternative.

COMMUNICATION BY SATELLITES

With the growing use of radio communications and fiber optics, the future for satellite communications does not look as rosy as it once did. Communications satellites are now used extensively for long distance telephone and video transmission. With fiber optics in place, use of satellites for long-distance telephone traffic will be reduced. Satellites suffer from signal delays, which makes them somewhat undesirable for phone conversations. The signal must travel up to a satellite, 22,300 miles high, and then back to earth. Even at the speed of light, that takes a quarter of a second. A person saying hello on one end will not hear a hello back for at least half a second, meaning he might go on talking during the silence. People talking over satellite connections thus always interrupt one another.

However, satellites do and will continue to have important uses in communications. Fiber optics cannot be used for broadcast applications. But a single beam from a satellite can extend over a whole continent, allowing virtually everyone in a nation to be reached at once by a signal sent from a common point.

It is for this point-to-multipoint communications that satellites will serve a vital role. The nationwide paging mentioned above, for instance, could not exist were it not for satellites.

Nor could the cable television programming networks, such as Home Box Office, exist. They use satellites to distribute their programs to local cable television companies. The major television networks are starting to use them in place of special telephone and radio links to distribute programs to their local affiliates.

Satellites also allow for point-to-multipoint teleconferencing. Companies, for instance, hold teleconferences to announce new products. The announcement is made in a studio, and those attending around the nation gather in buildings equipped with satellite receiving antennas. Universities also use this technique to broadcast lectures to other schools.

The next use of satellites might be to broadcast programs and information directly to homes and businesses from space. Currently, large dish-shaped antennas, at least ten feet in diameter, are required to receive the signals from today's communications satellites. Enough people have purchased such antennas that pay television services have begun to scramble their signals to prevent reception by people who don't pay. While some individuals use them to snatch programs directly from space, the size and expense of these antennas restrict their use. But new developments, including high-powered satellites, better receivers and more sophisticated transmission techniques, promise to soon allow individuals to communicate directly with satellites.

One possible use would be for distribution of television programs. Direct broadcast satellites (DBS), as the system is called, would allow television shows to be received in remote areas that are out of the range of conventional stations. In Japan, an experimental direct broadcast satellite in 1984 brought NHK broadcasts to remote islands for the first time. DBS could also bring pay-television services to areas not served by cable television.

While DBS is technically feasible, however, it is not yet clear how popular it will become, especially in the United States and Japan, which already have ample television programming. In the United States, most of the major companies that at one time said they would enter the business, such as CBS, the Communications Satellite Corporation and Western Union, have pulled back on their plans, because of high expenses and an uncertain market. One commercial venture that actually started, United Satellite Communications Inc., collapsed quickly. Expecting to reach 1 million customers within three years, it had only attracted 10,000 after one year in business. Cost might have been one factor. The service cost $39.95 a month for five channels of programming, plus a $300 antenna installation fee.

Satellite communications might also be used for mobile communications. The advantage over other forms, such as cellular technology, is that communications can extend to rural areas as well as cities.

As early as the mid-1970s, the National Aeronautics and Space Administration (NASA) demonstrated the ability to communicate via satellite with devices no bigger than walkie-talkies. In one experiment, ambulances in the South were equipped with systems that would enable a patient's medical signs to be transmitted, via satellite, to the hospital. The hospital could then send instructions for treatment back to the ambulance.

Several companies have proposed beginning such mobile satellite services, and it is expected such service will start in the early 1990s.

One company, Geostar Corporation, has proposed an even more futuristic plan, which would allow consumers to send brief messages to others, via satellite, using a small device. The system, by using several satellites and measuring the difference in the time it takes to receive the signals from the ground, could also determine a person's location. Geostar envisions that a person being mugged on the street could press a transmitter and send a signal to the satellites, which would determine his position and then summon the police. The proposal sounds farfetched and might never get off the ground, but it is an example of the type of thing that can be done via direct connections to satellites. Auto and trucking companies are also now looking at satellite-based navigation systems. This would allow, say, a trucking company to know at a glance where all its trucks are, to improve routing. Many of these systems would use satellite locators, perhaps the existing network of Loran navigation satellites or a new military positioning satellite system being placed in orbit.

In one system being studied by Chrysler Corporation, maps are stored on optical video disks and can be displayed on a screen in the car. Satellite signals determine the car's position and that is indicated on the map, so that the driver always knows his location.

THE NEW AGE OF TELEVISION

Television is also going through a major change. At first, television, as well as radio, was a passive media. A program was broadcast to everyone and was simply watched and listened to. There was little choice on the part of the viewer and little interaction with the programming.

In television's second era there are new sources of programming and new technologies to turn the television into more than what has been termed a boob tube. The television set is becoming a central command post of the home information and entertainment center.

One big change in television is the switch from analog to digital. This should come as little surprise, since this is a major trend in telecommunications. It will take some time before transmission of television is done using digital pulses instead of analog waves. But the processing of the signals inside the television itself is already starting to be done digitally. The first digital television sets began appearing on the market in 1984.

Digital technology allows the television to take on some of the characteristics of computers, producing, in effect, a "smart" television. The digital circuitry, for instance, can perform quick calculations to eliminate interference patterns known as ghosts. More intriging is the ability to freeze a picture frame electronically in the same way a computer stores data. Some Japanese companies have already developed printers that can print out individual frames from a television screen. Digital television sets, moreover, can display more than one program at a time. The first digital set on the market, sold by the Matsushita Electric Industrial Company, allows a second program to be seen in a small window inset on the main screen. One can watch a television show, for instance, and in the little window on the screen, monitor children playing in the next room and seen by a video camera. Some digital television prototypes were able to display nine programs at once, although action was in slow motion in eight of the pictures. Digital technology also makes the television more suitable as a computer display. Finally, it offers the potential for making one television set capable of receiving broadcasts under any system. Currently, there are four incompatible broadcast standards used in different parts of the world. A television set purchased for use in the United States, for instance, cannot be used in Europe.

Another improvement is multichannel or stereo sound, which

is already popular in Japan and is starting to be used in the United States on shows such as "Miami Vice." In addition to improving the sound for musical shows, the systems can allow for two separate soundtracks. The same programs can be broadcast in two different languages. Other innovative uses can be made of the technology as well. On some baseball telecasts in Japan, viewers are given the choice of announcers — one impartial and one favoring the home team. When traditional Kabuki theater is televised, the second audio channel is used for detailed commentary.

Also coming is high-definition television (HDTV). In this system, picture resolution is at least twice as high as in current broadcasts. HDTV is not likely to be used for general television broadcasts for quite a while because the frequencies are not available. If conventional broadcast frequencies were devoted to high-definition broadcasts, there could only be one fourth as many stations. And the existing television sets would no longer work. HDTV also suffered a setback in 1986, when an international committee meeting in Yugoslavia failed to come to an agreement on standards for the new system.

However, in the future, cable, optical fiber and satellites might be used for high definition television. Some think broadcast high-definition televison will use new UHF frequencies and coexist with existing television, much as superior FM radio came in and now coexists with AM radio on different channels. Others think high-definition broadcasts will succeed only if they are compatible with existing television sets. Under these schemes, an existing set would still be able to receive high-definition broadcasts, though with low resolution, in much the same way as a black-and-white set can receive color broadcasts, but only in black and white.

High-definition television is more than a mere improvement in quality. It would make video images as high in quality as 35mm film. This would allow for video to take the place of photography. Some movie producers are using high-definition videotape in place of film, because videotape does not have to be developed and can be easily edited. A few movie theaters are experimenting with huge video systems to supplement or replace movies. That would allow the theaters to show closed circuit television of events such as boxing matches. With conventional television, the picture becomes blurry with screens larger than twenty-five inches or so, but that is no problem with high-definition television.

CABLE AND PAY TELEVISION

Cable television was originally begun to improve reception of broadcast signals in remote areas. Now it is becoming a major form of communications in and of itself. Cable frees television from the handicap of limited available transmission frequencies, allowing for many channels to be broadcast. Cable prophets envisioned that rather than broadcasting, cable would open up an era of "narrowcasting" in which programs could be tailored to narrow interests. So far, however, the promise has not materialized, because programming is expensive. A huge base of viewers is needed to justify the costs of producing a program. Indeed, ambitious plans for cable in the United States and Britain are being scaled back.

One way of serving more narrow audiences is to have people pay for programs. This is being done by pay-cable services and over-the-air pay television. These take advantage of electronic boxes, called decoders, that allow some sets to receive the programs. Most pay services charge a flat monthly fee. But one concept that is gaining favor is pay-per-view, in which viewers pay to see particular events. But the events are offered to everyone at the same time, just as a movie theater offers only a few showings a day at specific times. The next step would be to allow viewers to request what they want to see whenever they want to see it. This would be the video equivalent of a data bank. These are not likely to arise until far into the future because there is not enough bandwidth available, even on cable, to serve many customers with different programs at the same time. But such a service might arise when most homes are served by optical-fiber phone lines.

INTERACTIVE TELEVISION

So far, television has largely been a passive activity. A person merely watches. But technology also will allow people to interact more with what they see.

One potential technology in this vein is teletext. In this service, data and text are broadcast to viewers. Often the data are sent in what is known as the vertical-blanking interval, the scanning lines that separate one picture frame from the next. This interval is the black bar that appears on the screen when the picture is not aligned vertically.

Teletext is similar to videotex, which is delivered over telephone lines. The difference is that videotex is a two-way system, meaning users can request specific information and conduct transactions, such as paying bills. Teletext is a one-way system, meaning viewers can only read information and choose from what is offered. Usually, a few hundred still frames of information consisting of text and crude graphics are broadcast in a cycle, and the viewer, using a keypad, chooses what frames he wants to see. The advantage is that it is envisioned that teletext would be provided free of charge.

So far, however, teletext has had poor results, except in Britain. In the United States, all systems that have started have done poorly, partly because of the high cost of decoders. Users have other ways of obtaining information on the weather, sports scores and the like. One way teletext might catch on is by using the service to supplement television shows and commercials. A commercial for a vacuum cleaner might advise the viewer to turn to a certain teletext page to receive a list of local dealers. Or, during a baseball game, a teletext page might provide statistics on the current batter.

Even more disappointing than teletext has been the experience with two-way television. This technology was once envisioned as enabling people to "talk back to their television sets." The forerunner of such systems was Warner-Amex's Qube. This allowed viewers to respond to questions on the air using home decoders. Instant polls could be conducted. Viewers could answer quiz show questions at home. And, in one experiment, home viewers called the plays in a semiprofessional football game and watched the teams on the field execute the plays.

While intriguing, however, Qube, which was tried in Columbus, Ohio, and several other cities, never really was worth the cost involved. Many critics condemned it as gimmickry, as a device to help Warner-Amex win cable franchises. By 1983, Warner-Amex had run into severe financial problems and was drastically cutting back on its plans for Qube.

Still, interactive television is a dream that remains alive. One new form could involve playing games linked to television programs. In 1986, several companies applied for permission from the Federal Communications Commission for such games. In one version, for instance, inaudible tones broadcast with a particular show would direct the movement of toy robots on the floor, so that a child not only would watch the show, but would get involved in the action.

Yet another form of interactive video could be provided by optical disks. Already, video disks, linked to a computer, can allow a viewer to switch instantly from one video segment to another. In an entertainment program, for instance, a viewer could make choices that would change the plot as the program went along. Such disks, however, are more often used for training than entertainment. A new technology developed at RCA would allow digital data, including video, to be stored on a compact audio disk. One possible use could be for so-called surrogate travel. A user can take a tour of a town or a museum on the video screen using a joystick or control panel. As the user directs the stick to move forward or turn left or back up, the disk would automatically provide the proper picture to make it look to the user as if he were really there.

PERSON-TO-PERSON VIDEO

Video communications can also be used for person-to-person communications through a video conference, or videophone. Today this is extremely expensive. Special high bandwidth lines or satellite circuits are needed to transmit the video. A special room, which can cost several hundred thousand dollars, is usually needed, equipped with microphone, cameras and viewing screens. Video conferences have been used successfully for point-to-multipoint communications, such as a worldwide press conference, where a carefully rehearsed presentation is made. Many corporations now have their own television studies and satellite networks linking their various offices. These can be used for internal announcements and training. But only a few big companies so far use teleconferencing between two groups. Moreover, people who are not used to being on television also find it somewhat awkward to conduct such meetings.

That could change, however, as people get more used to the idea. In addition, data compression techniques have been developed that reduce the need for special transmission lines. It is now possible to send full motion video at a rate of 56,000 bits per second, a level that can be handled by an ordinary phone line. Also, portable units are being developed that can be wheeled from room to room, doing away with the need for a special conference room. Still, it is not known if social customs will change enough to make videophoning popular. As one analyst noted, "I wouldn't want anyone to see my messy desk."

As a cheaper alternative to video conferencing, some companies are using slow-scan video. In this technique, a series of still frames, like a slide show, is transmitted, rather than a moving video picture.

Indeed, one big use for video technology in the future might be to replace photography. Sony in 1981 announced its Mavica, a video still-camera that stored images on a magnetic disk, rather than on film. One advantage was that the pictures could be instantly seen on a television set, without the need for developing. Another advantage was that the images could be instantly transmitted electronically over the telephone line, just like any other data. That would allow Mom and Dad to take a picture of their new baby and instantly transmit it to Grandma, who would display it on her television screen.

The Sony system was given a dry run during the 1984 summer Olympics. By using it, the *Asahi Shimbun,* one of Japan's largest newspapers, was able to get the latest photos into the paper in time for deadline. The rival *Yomiuri Shimbun* tested a similar but even superior system developed by Canon.

Despite the success of the early tests, however, the video still cameras have not come to market as fast as the developers initially predicted. The image quality is still inferior to that of photographs. And the expense is so high that the product is likely to appear as a commercial product, not a consumer product, for several more years.

Another means of image transmission, which has long been in use, is facsimile transmission. This form of communications is much more important in Japan than in the United States and Europe, because it can handle the Japanese characters that are difficult to type. Facsimile transmission has undergone its own revolution. Once it took as long as six minutes to transmit a page. Now it takes less than twenty-five seconds over a telephone line. With special high-speed digital lines, it can be done in a few seconds. The same computer technology that might one day be applied to voice is already being applied to facsimile transmission. Facsimile networks can store a document for later delivery, or process the information en route to change its size, or even make multiple copies of the same document for different destinations. In a recent development, some companies are coming up with systems that allow personal computers with optical scanners to serve as facsimile machines. Still, facsimile-based transmission systems have not been successful businesses yet. Federal Express, known for its overnight document delivery service, which is a big success, tried a service called Zapmail that would deliver documents within hours. Federal Express would pick up the document and fax it to its office in another city, where it would be printed and delivered. But the system was abandoned in 1986 after losing hundreds of millions of dollars.

THE BIG ISSUES

The new telecommunications will indeed have a vast effect on the way in which we all live. But the effect will not always be a positive one. The whole social fabric of society is likely to be affected, partly in ways that cannot be foreseen now. But it is unlikely that the future will indeed become a "teletopia," as Japanese telecommunications planners have envisioned. Numerous serious issues must still be dealt with, and numerous questions remain. What follows are some of the concerns.

HAVES AND HAVE-NOTS

If communications lines are indeed becoming the highways of the next century, then lack of access to communications could be as much a handicap in the future as not having a car is today.

This brings up the issue of whether society will be fragmented in the future into information haves and have-nots. Until now, telecommunications capability has been fairly widely distributed, at least in the United States. Radio and television have been free and available to all who can afford the receiving equipment, which is virtually everybody. Telephone service throughout the world has been run or regulated by the government. In the United States, one goal of government regulation was to provide universal telephone service. Rates were held down to allow everyone to afford basic telephone service. The revenue was made up by charging higher for certain services, such as long-distance telephone calls, used mainly by businesses and wealthy individuals.

But the old government regulations could not keep pace with the era of rapid innovation, so the old system is falling apart at the seams. The United States has largely deregulated telephone service and ended the monopoly held by the American Telephone and Telegraph Company. Similar steps at ending the government monopoly are under way in Japan, Great Britain and Canada.

Competition means that those who can pay for the most sophisticated technologies — meaning businesses and the wealthy — will attract better service and more options. Meanwhile, service for the poor will diminish because no company can afford to subsidize service to the poor by charging the wealthy more. In an era of competition, the wealthy would move on to a new supplier if the supplier they are using overcharges them to subsidize other customers.

Already, in the United States, local telephone rates are rising. Electronic mail, largely available to businesses and wealthy individuals with computers, is taking business away from the postal service, which offered cheap, albeit somewhat slow, delivery to all. If the postal service loses too many prime customers, it will have to make up for it either by charging more to the poorer customers who have no alternatives, or by letting service deteriorate. An international debate is raging over whether competition should be allowed in international satellite communications, an area that is now the sole province of Intelsat, a multinational organization. Business users in the advanced countries want competition to provide them better service. But poorer nations fear losing their low-cost service and being left further behind. Finally, pay-television services in the United States are undermining network television. Some experts are already predicting that by 1992, the Olympics, an event meant to draw the people of the world together, will be available in the United States only on pay television, which will be able to outbid the networks. The danger for information haves and have-nots, therefore, is clearly present and promises to spark numerous controversies in the years ahead.

CULTURAL IMPERIALISM
AND SOCIAL COHESIVENESS

Telecommunications has already done a great deal to make the world a smaller place. Television and movies have helped spread the American culture far and wide. Columbia University students are studying the Soviet Union by watching television programs intercepted from the Soviet satellites. The future improvements in telecommunications promise to do this even more. Satellites, in effect, make it possible to have international networks, instead of merely nationwide television networks.

But this can also be undesirable. One fear is that the nation that controls the information and telecommunications technology can impose its views and its culture on other nations. Already, lesser developed nations are concerned that the world sees itself through the eyes of a handful of Western news organizations. The American government, already using radio and television to get its views across to other nations, is now harnessing the satellite. The United States Information Agency set up Worldnet, a satellite teleconferencing system that will allow officials of the federal government to hold press conferences with reporters around the world.

In addition to political power, satellites promise to lead to more controversies regarding "cultural imperialism." Canada, worried that it is not developing enough domestic television, has cracked down on the use of satellite earth stations, which are being used to pick up American programs sent by satellite. When direct broadcast satellites come to Europe, the satellite beam will spill over into many nations with different languages and cultures.

The new telecommunications technology can also work to fracture social cohesiveness. Americans, for instance, to some extent have a common culture because they all watch the same television shows. But with new communications alternatives, with the prospect of information haves and have-nots, this common thread could disappear. If narrowcasting catches on instead of broadcasting, people will no longer all be watching the same shows. Similarly, with newspapers and magazines, everyone is exposed to common fare. But with an electronic data base, people can retrieve only the information of special interest to them. A person reading a newspaper might be interested only in sports but might be drawn to an article on a significant political development. With a customized electronic newspaper, the person would see only articles on sports.

"These trends all point to the same lugubrious irony," wrote Benjamin R. Barber, professor of political science at Rutgers University. "The new telecommunications technology may destroy the unified democratic culture that was the greatest civic achievement of television's first age, without realizing any potentialities that are the greatest civic promise of its second age."

ELECTRONIC SURVEILLANCE AND PRIVACY

The story is told that the Central Intelligence Agency once gathered together a bunch of experts and asked them to design the most efficient system by which the Soviet Union, or any other nation, could monitor the whereabouts of its citizens. After working on it for several days, the group came up with its answer — an electronic funds transfer system.

These systems, as discussed earlier, would allow a person to pay for purchases by electronically transferring money from his account to the store's account. But every time such a transaction is made, a little electronic record is made. It is thus possible to know that John Smith purchased a book at 3 p.m. on June 25 at a particular book store. It might even be possible to know what kind of book it is. What if it were a pornographic book, or a book of political ideas viewed as subversive by the prevailing regime?

The threat is clear. Electronic transaction systems pose a threat to privacy. Two-way cable and videotex systems record what information a person wants and what shows he watches. More advanced electronic funds transfer systems can track a person around the city and around the nation. Laws and technological safeguards must be put in place to ensure that such information is not used wrongly, such as for blackmail or surveillance. Otherwise, the teletopia might turn into what 1984 was supposed to have been.

Chapter Six

The Earth
And Its New System of Satellites

By Wayne Biddle

Wayne Biddle is the author of Coming to Terms, *a book on scientific terminology, and a number of magazine articles on science and government.*

Sometime in the deep past, long before there were wheels, a feral poet probably realized that the moon goes around the earth. This dreamer couldn't prove it, of course, and he or she was surely too frightened to confide the matter in anyone else. But before going mad with the knowledge, the poor creature must have seen itself up there riding around and around, smiling at its friends and screaming at its enemies, holding the fundamental force of the universe by the tail.

Today, the smiling and the screaming have both assumed dimensions even a genius could not have foreseen. There are more than 5,000 man-made objects in orbit, placed there by more than a dozen nations. If there is one aspect of space exploration that has become utterly commonplace, it is artificial satellites. They opened the space age in the late 1950s and, unless we are careful, they may be the final instrument of our demise. Whereas manned space travel has turned out to be an economic burden too heavy in all but the most restrictive forms, unmanned satellites are a mature industry with enormous technical vitality.

Conclusive evidence of this maturity came several years ago, when the Federal Communications Commission deregulated the nation's commercial satellite operators. The decision meant that such companies as RCA, American Communications Inc., Western Union Corporation and Satellite Business Systems are free to raise or lower their rates, offer new services or discontinue old ones without government approval.

"If customers need capacity, all the carriers have it," said Jonathan Miller, managing editor of an industry trade publication. "There simply is no longer any market monopoly in this industry."

So far, the three most important things that satellites can do are look at the earth, relay information and carry weapons. For these tasks, they are reliable and affordable, though international treaties have outlawed the third capability.

There are other tasks that satellites can perform, mainly in the area of scientific research. But government budget priorities in recent years that favor military space applications mean that these promising realms will remain largely unexplored for the immediate future. Placing various sorts of telescopes in orbit, for example, can provide revolutionary new views of the universe through what scientists call a "clean window," but the hardware is impossible to obtain without large infusions of tax dollars.

There is something else that satellites might be able to accomplish, though the technical community is still divided over whether it's worth devoting scarce funds to find out. Environmental conditions beyond the earth's atmosphere — high vacuum, extreme heat or cold, low gravity — could enhance certain manufacturing processes. Whether the advantages outweigh the cost of reaching orbit and maintaining equipment there has been an issue of debate for years. The American space shuttle, in reality an expensive developmental vehicle that is feeling the way for more practical space trucks of the future, has already served as a federally supported test center for commercial experiments in space manufacturing. But there is still a great deal of skepticism about such work outside National Aeronautics and Space Administration circles.

Gazing through a crystal ball at the future of satellites, therefore, reveals a spectrum of possibilities that range from the nearly certain through fantasy. For the sake of engineering realism, it is perhaps best to avoid speculating about such things as factory-moons that will house a hundred pharmaceutical workers (or their robot equivalents), laser guns that will vaporize enemy missiles in flight, and manned observatories where members of college astronomy departments can spend their sabbaticals. Two broad categories of satellite applications will suffice: reconnaissance and manufacturing.

RECONNAISSANCE

Reconnaissance, which can be for either military or civilian purposes, takes into account all aspects of how satellites can "look" at the earth. Looking, in this regard, may be thought of not only in terms of gathering images, but as a way of collecting many types of information about the earth's surface and atmosphere. Images, of course, can be constructed from many parameters besides visually translatable data — temperature, for example, or radioactivity. "Remote sensing" is perhaps a better overall term here, though it smacks of jargon.

Owing to the pressure of military demand, the cameras and radars commonly used in imaging reconnaissance satellites have reached a high state of art. An entire deskbound technical field, known as image analysis or photointerpretation, has evolved to support spacebound hardware. The future of reconnaissance satellites is thus tied as much to advances in computer-aided image enhancement as to satellite hardware.

In a world where a popular mystique has grown up around the notion of spy-in-the-sky satellites that can purportedly read automobile license plates from space or discern the faces of world leaders as they step from limousines, it should be kept in mind that there is no magic technology held exclusively by the military. (Atmospheric turbulence creates a resolution limit of about ten centimeters, though image digitalization by computer can fill in the blurry parts, so to speak.) The civilian Landsat and Spot satellites, for example, are capable of providing excellent images of a quality that Air Force or Central Intelligence Agency satellites do not fantastically improve upon. What the Pentagon and the intelligence agencies possess to advantage is a vast support network, both human and cybernetic, for enhancing the information received from space. Their satellites, moreover, can handle more information faster than civilian cousins.

But even this advantage is not absolute. The Geophysical Institute of the University of Alaska in Fairbanks administers a Landsat Quick Look facility as a source of data for users who want access to enhanced images within hours after the satellite passes over a target area. In fact, enhanced and enlarged views of a section of Arctic Ocean ice obtained by Landsat were used by American analysts to surmise that the Soviet Union has been conducting tests of how to launch missiles from submarines under the polar ice cap.

There is no technical reason, in other words, why instantaneous "real time" broadcasting of earth images now available to military and intelligence officers cannot become available for commercial use in the future. The utility of such technolgy for worldwide disaster relief, aircraft and automobile traffic control in extended metropolitan areas such as the Boston–Washington corridor, or scientific observation of remote regions, is obvious. A small network of intelligence satellites of the most advanced type, such as the CIA's KH-11 series, is believed to be worth billions of dollars, yet something as mundane as a traffic jam can easily cost as much to a city's economy if repeated often enough at morning rush hour.

According to William J. Perry, Undersecretary of Defense for Research and Engineering in the Carter Administration, the supersecret satellites used to gather military photographic intelligence have been rising in altitude in recent years. This is a development that bodes well for eventual commercial applications, since it means the expensive satellites will not have to be replaced as often. When first sent aloft in the 1960s, they were confined to orbits about 100 miles high in order to provide the best images possible at that time. At such low altitudes, where the drag of the atmosphere is still significant, their lifetimes were measured in months, sometimes only in weeks for the largest and heaviest cameras. The primary reason why the Soviet Union has launched many more military satellites than the United States is that it has had to replace its spy satellites much more often because they were not as miniaturized and lightweight as analogous American hardware.

As with so many other endeavors, the availability of microcomputers is helping to push reconnaissance satellites beyond the limits of their present ability. Images detected by their cameras, whether with visible light or other portions of the electromagnetic spectrum, are enhanced by computer techniques. Roughly speaking, what the cameras cannot resolve the computer can supply. If the camera snatches only a portion of the number five on the deck of an aircraft carrier from an altitude of 1,000 miles, for example, a computer can fill in the missing pattern, either by recognizing what the fragment belongs to or by asking for the opinions of other orbiting cameras looking at the ship from different angles and spectrums.

All that is required to do this in real time — that is, simultaneously vis-à-vis the remote viewer — is sufficient on-board computing power and generous transmitting room to converse with neighbor satellites. The tracking and data relay satellites (TDRS) placed in low-earth orbit by the space shuttle are a first step in providing the transmission capacity.

As the capacity grows, the applications get sexier: three-dimensional pictures, for example, cinema quality two-dimensional television, or individualized reception on wristwatch devices à la Dick Tracy. The latter possibility brings up the problem of making power levels high enough to excite very small antennas. These two basic requirements, high data capacity and high power, could bring new approaches to satellite communications by the beginning of the next century, including orbital space platforms and antenna farms.

If there is enough demand for huge platforms containing many elements, it might be cost-effective to manufacture them on the moon. Solar-energy satellites in high orbits could be used to power and stabilize reconnaissance and communication devices at lower altitudes.

Real-time military reconnaissance and communication is of enormous appeal, needless to say, though real-life episodes such as the American attempt in 1980 to rescue embassy personnel held hostage in Iran have shown that instantaneous knowledge of distant events is not necessarily helpful. In the name of futurism, therefore, it might be more useful to envision possible nonmilitary applications.

Imagine, say, a natural disaster such as another great San Francisco earthquake, which is presumed to be inevitable and due quite soon. Imagine the loss of local electric power, TV, radio and transportation. In the face of chaos even with the best emergency preparedness, the availability of all-weather reconnaissance satellites to provide rapid damage assessment at the command of local officials would be of inestimable value. From a prepositioned post near the quake zone, relief workers could quickly determine where to channel scarce resources. They would know — with a visual resolution near ten centimeters — conditions anywhere in the disaster area.

Imagine a similar system devoted to flood or drought control. Not all of the orbiting reconnaissance satellites would have to be devoted to active earth-watching. Some might be relatively simple "alarm sensors" tuned to detect changes, such as color or density of vegetation, in fragile environments. The value of such a system is approximated by meteorological satellites currently in geosynchronous orbit, which provide a few more precious hours of predictive ability for weather forecasters.

In agriculture, environmental surveying, city planning, geological studies and land use, reconnaissance satellites are expected to provide an economical means of management and decision-making in the next century.

MANUFACTURING

Besides reconnaissance and communications, which are relatively mature industries, the only other commercial use of space that is not entirely in the realm of futurism is materials processing.

Theoretically, any manufacturing process in which gravity causes unwanted perturbations could be a prime candidate for space-based industry. Experiments in a 700-pound laboratory carried in the payload of a shuttle mission showed, for example, that pharmaceuticals could be processed in the gravity-free environment of space. Conducting such work is costly. Often aspects of the space environment can be duplicated on earth. And a National Research Council panel concluded in 1978 that even when gravity has an adverse effect on a process, "stratagems for dealing with it can usually be found on earth that are much easier and less expensive than recourse to space flight."

Still, processes that might benefit from a low-gravity environment would include the growing of semiconductor crystals, the fabrication of precision machine parts, containerless reactions (particularly the preparation of high-purity ceramics), elimination of the settling of particles for substantial periods of time, purification and myriad other chemical reactions. But economics is the ultimate criterion in space, not theory. If a certain space-based manufacturing process is as commercially superior to its earthbound alternative as Telstar was to transatlantic telephone cables in 1962, then it will eventually become as pervasive as communications satellites are today.

The NRC panel said that "there is oportunity for meaningful science and technology developed from experiments in space provided that problems proposed for investigation in space have from the outset a sound base in terrestrial science or technology and that the proposed experiments address scientific or technical problems and are not motivated primarily to take advantage of flight opportunities or capabilities of space facilities."

The panel noted that it had not discovered any examples of economically justifiable processes for producing materials in space, though there were areas worth exploring in the "preparation of specialized exemplary materials." Recent experience on the space shuttle shows that even this limited work has a long way to go before it becomes self-sustaining. The panel's conclusion on electrophoresis seems prophetic: "The objective of learning more about how electrophoresis apparatus should be designed and how gravity may affect the electrophoretic process will best be answered through well-planned terrestrial research rather than experiments in a low-gravity environment."

REPAIR WORK IN SPACE

As satellites become larger and more expensive, the ability to repair them in space may become desirable. For the foreseeable future, repair work will be limited to low-earth orbits, as demonstrated in primitive operations of the space shuttle so far. Only the most expensive satellites will ever be candidates for space-servicing, of course — those whose intrinsic value is significantly greater than the cost of sending up and bringing back a repair crew.

To make such servicing possible, satellite design will have to be changed. It will never be feasible to dispatch a complete test laboratory with a full complement of technicians to troubleshoot an errant satellite in situ. But it may be attractive as well as practical to replace faulty plug-in modules whose failure has been announced by an on-board computer. So far, however, satellites have not been designed or constructed with service visits in mind.

As long as the maintenance of human life-support systems in space remains formidably expensive — and there is no reason to believe it will ever be otherwise — there will be little justification from an engineering perspective for satellite repair by human hands. In most instances, it will be far cheaper to design several layers of redundancy into critical components than to depend on the arrival of human troubleshooters.

Expensive satellites that consume inexpensive materials, such as orbital positioning fuel or photographic film, might be an exception to this rule of thumb. If it were not for the essentially limitless budgets available to the Air Force and CIA, which allow them to replace high-priced reconnaissance satellites at short intervals, perhaps space-servicing would already be commonplace. There are still too many technical alternatives, apparently, to the risky and burdensome task of manned visits.

SPACE WEAPONS

Although international treaties currently ban the placement

of nuclear weapons in space, there is enormous momentum — measured in terms of fiscal dollars and research manpower in both East and West — to develop nuclear and conventional orbital weaponry. History does not provide much hope for those who would seek to curtail such extensive technological efforts. Any vision of the future should take into account the likelihood that some types of belligerent devices will be present in space.

Until recently, the superpowers have found economic and technical support for five broad military missions in space that do not involve shooting at anyone — communications, reconnaissance, navigation, meteorology and geodesy. Although other activities, such as antisatellite weapons and ballistic missile defense, are attracting far more attention, these five traditional missions will probably remain the most important from a military perspective in the near future.

In the realm of military satellite communications, a great deal of research effort is devoted to using lasers as a transmitting medium, deploying small mobile receivers on the ground and devising ways to protect satellites from ground-based jammers. In reconnaissance, there are proposals to place sensors in orbit for detecting enemy bombers or cruise missiles, to use satellites instead of inertial navigation systems for guiding ballistic missiles, to put cameras in space for surveying enemy territory after a nuclear strike and to devise space-based technology for finding enemy submarines as well as communicating with our own subs.

President Reagan's desire for an antimissile system in space, using lasers or other directed-energy weapons, has already received intense scrutiny by the press and diverse elements of the scientific community. The consensus even among Pentagon scientists is that such a system is years away, assuming Congress would continue to fund requisite research. By their own admission, too, the work will require fundamental scientific breakthroughs. This is no Project Apollo, in other words, nor does it compare closely with the development of the atomic bomb.

But there are many other potential weapons applications in space that are perhaps more readily attainable. For example, several aerospace companies are at work on an "orbital transfer vehicle," or spaceplane, that could be used for attacking enemy satellites. Conventionally armed "space mines" — bombs that explode when a threatening object comes into their vicinity — may already be deployed around critical military satellites. But physical violence is not absolutely necessary for threatening a satellite. Myriad passive jamming techniques are available and probably preferable.

The key to future weapons applications in space, as demonstrated by Soviet and American antisatellite systems already in existence, is to adapt conventional explosives or nonexplosive direct-impact devices for use in orbit. Because of the electromagnetic havoc produced by fission and fusion explosions, nuclear bombs may in fact have no place in space, where one's own fragile electronic equipment would be just as vulnerable to pulses of subatomic particles as the enemy's.

Like the future of automobiles, the future of satellites is mercifully colored by long experience in the past. Just as we know by now that cars are never going to look like the hyperstylized bubbles envisioned in the 1930s, it is safe to say that satellites will never assume the scale of artificial moons predicted by some science writers.

Satellites have been good business for two decades. As such, they are not susceptible to the futuristic dreaming that plagues other endeavors in space. The opinion of satellite businessmen, whether in commerce or the military, is that satellites will largely continue to do what they already do well. Great leaps forward will come in the name of market economy rather than under the flag of exploration.

But what about the exotic possibilities? Will there ever be space colonies, personalized communication links or microchip factories in orbit? With the richest nation on earth no longer able or willing to lavish money on space spectaculars, at least of the nonmilitary variety, it would seem that from a strictly economic point of view these vast projects are unlikely. Only a major reordering of political priorities could lead to orbital space colonies, for example, that require a commitment of resources now almost beyond imagination. This should not be cause for disappointment among space buffs, however — merely a sign that many dreams have already come true and are now generating their own workaday world.

With apologies to Konstantin Tsiolkovsky, the Russian rocketry and space science pioneer: Mankind may not remain forever on the earth, but, in a quest for light and space, will at least conquer the whole of low-earth orbit.

Chapter Seven

The Exploration of Space

By John Noble Wilford

John Noble Wilford, science and space writer for The New York Times, *is the author of* We Reach the Moon, The Mapmakers, The Riddle of the Dinosaur *and a coauthor of* Spaceliner: The Story of the Space Shuttle.

Out beyond the most distant planets, out toward the fringes of the solar system, a hardy little spacecraft cruises on and on into the unexplored. No machine of human design has ever gone so far. And still it sends faint messages billions of miles back to Earth every day, whispers of discovery. Someday the eight-watt radio transmitter will go silent, but the little craft, Pioneer 10, will cruise on and on into the greater unknown, the first man-made object to leave the solar system and head out toward the stars.

The travels of Pioneer 10 symbolize the past and future of space exploration. Humans and their machines have reached out to incredible distances in the first quarter-century of spacefaring. They have been to other worlds and had revealing glimpses deep into the universe. But all this is only a beginning.

When Pioneer 10 was launched March 3, 1972, from Cape

Canaveral, Florida, no spacecraft had ventured farther than Mars. The Apollo astronauts had visited the moon, and the unmanned Mariners had flown by and photographed Venus and Mars. Hundreds of spacecraft had orbited Earth. Pioneer represented a major extension in the range of space exploration. The 570-pound craft made its way safely through the asteroid belt, a region littered with rocky debris between Mars and Jupiter. It flew within 81,000 miles of Jupiter's cloudtops on December 2, 1973, returning the first close-up images of the sun's largest planet. Still Pioneer kept going, across the orbits of Saturn, Uranus, Pluto and Neptune. On June 13, 1983, Pioneer became the first spacecraft to depart the realm of the known planets.

Scientists with the patience to decipher Pioneer's radio messages are learning for the first time what it is like in the outermost solar system. It is cold and dark and empty, as they knew it must be. A tenuous wind of solar particles, the million-mile-an-hour solar wind, still blows outward. Cosmic rays race inward. A virtual vacuum it may be, but nothingness, it seems, is a relative condition.

If the spacecraft survives long enough and the scientists are clever enough, more exciting discoveries could lie ahead for Pioneer 10. It might be able to detect gravity waves, which were theorized but have never been detected. It might locate the source of the mysterious force tugging at Uranus and Neptune, a gravitational force suggesting the presence of some as yet unseen object — perhaps the long-sought Planet X or a dim companion star to the sun. The spacecraft may also function long enough to report back the answer to the question: Where does the solar system end and interstellar space begin?

Someday, of course, even the durable Pioneer 10 will lose touch with those who sent it off on its long journey. The radio will go dead. Guidance sensors will lose sight of the sun. Even then, perhaps in the next few years, Pioneer 10 will cruise on, an artifact of mankind's first bold thrust beyond the solar system.

With this in mind, some scientists have contemplated Pioneer's infinite and eternal future. They calculate, for example, that 10,506 years from now Pioneer should have its first fairly close encounter with another star. That is when it will pass within 3.8

light-years of Barnard's star, a small, cool star. Then it will be on to Ross 248 some 32,000 years from now. Also, in the next 862,063 years, by current calculations, Pioneer will aproach the vicinities of eight other stars, including Altair, a star bigger and nine times brighter than the sun. Somewhere along the way, the derelict craft could come close enough to a star system to be intercepted by intelligent beings, if there are any elsewhere. For this reason Pioneer carries a plaque with images of a man and a woman, a diagram of the solar system and other symbols that might help others locate the origin of the little craft.

Five billion years from now, the scientists further calculate, Pioneer should be wandering about the outer rim of the Milky Way galaxy, of which Earth and the sun are but a tiny part. Or were a part, as the case should be by then. The sun is expected to burn out and die in 5 billion years, and with it Earth and the other planets will vanish. Pioneer 10, therefore, will very likely outlast human life on Earth and Earth itself — which is why the little craft symbolizes not only the first steps in space exploration but also the promise of greater and more distant exploits in the years ahead.

Projecting the future in any endeavor is fraught with risks, but never more so than in the realm of space exploration. Timid minds fail to imagine all that can be. We have already far exceeded their modest forecasts. More imaginative minds fail sometimes to distinguish between what can be and, given economic and political realities, what is more likely to be. We have yet to realize their visions of luxurious space stations, lunar colonies and astronauts journeying to Mars. And tragedy comes along and reminds us again of the costs of venturing beyond Earth. The destruction of the space shuttle Challenger in January 1986, taking the lives of its crew of seven astronauts, brought the American space program to a virtual standstill. Projects and goals were questioned as never before. Public support for space exploration continued to be strong, but this was not being translated into adequate financial backing for the means to accomplish far-ranging ends. In the aftermath of the Challenger disaster, we realized that we had reached a point where our capabilities exceed our commitment to an expanded era of space exploration.

There were signs, however, that many nations of the world are beginning to think more boldly again about where they might go and what they might be doing in space in the next generation or two. The time is approaching when the United States and the Soviet Union, the main participants so far, will find the nations of Western Europe, Japan, China and others taking an increasingly active part in the human space enterprise.

SHUTTLES AND STATIONS

Even before the Challenger accident, the manned space programs of the two superpowers, the United States and the Soviet Union, were in a period of transition. They still are, but prospects for the near future are relatively easy to discern.

After the end of the Apollo lunar-landing era, in the early 1970s, the United States concentrated its efforts on developing a revolutionary new spaceship, the reusable space shuttle. Since the early 1970s, after dropping out of the so-called moon race, the Soviet Union concentrated on gaining experience in long-duration flights in Earth orbit, aiming toward what Soviet leaders see as a permanent occupation of orbit with large space stations. Despite these differences in approach, however, the efforts of the two countries are likely to converge in the next decade or so. The United States, which has a space shuttle, is planning a space station, and the Soviet Union, which has a space station in orbit, is developing a reusable space vehicle of its own.

Much more is known of the American plans, though they have been disrupted by the Challenger accident. When the shuttles fly again, they will be used to deploy some (but not all) large satellites for scientific studies, military reconnaissance and other purposes. Astronauts in the shuttles will be called upon to repair and service orbiting satellites. They will conduct experiments in manufacturing and materials processing that take advantage of the unique microgravity conditions of space. In the early 1990s, according to current schedules, a replacement for the Challenger should be ready, bringing the shuttle fleet back up to its full strength of four.

But never again will the shuttles be counted on as the nation's sole launching system. The Challenger disaster proved the policy of depending on a sole launching system to be unwise, to put it mildly. The Defense Department will have a complement of unmanned rockets in production to fill many of its satellite-launching needs and also to deploy payloads for the Strategic Defense Initiative, better known as the Star Wars program. There will soon be larger versions of the workhorse Titan 3 rockets. These Titan 4's will be able to carry many of the large-sized payloads that would have been totally dependent on the shuttles. In addition, a new line of Delta rockets will be hauling smaller satellites, and plans are under way to develop a heavy-lift rocket capable of carrying payloads even heavier than those designed now for the shuttles. Such rockets could prove invaluable in the next decade, if major components of Star Wars technology are ever to be deployed.

Still, the shuttles will be the only means for carrying American astronauts into space well into the next century. The shuttles will be essential for assembling and constructing large facilities in orbit, such as a space station. But no one now expects the current shuttles to realize their original goal, which was to reduce significantly the cost of going into space. The goal had eluded the shuttle program even before the accident, owing to funding delays and cost overruns and the difficulties in servicing the shuttles quickly enough to keep them flying with a brief ground turnaround time. Someday, perhaps in the early twenty-first century, smaller, more efficient shuttles may be in service. Engineers in the United States, Europe, Japan and the Soviet Union are developing concepts for an aerospace plane. This winged vehicle would be capable of taking off from a runway, climbing into orbit and then returning to runway landings. Being able to take off from runways would make these vehicles more versatile and less expensive to operate than the current shuttles. But the technological challenges are not inconsiderable.

The Russians will also keep building on the foundations they have laid since 1971. That was when the Soviet Union launched the first of its Salyuts, prototypes of space stations. Each Salyut weighed about twenty-one tons and was capable of holding crews of two to five people for long periods of time. In 1984, three cosmonauts returned to Earth after spending 237 days in Salyut 7, an endurance record for spaceflight. The Russians give every indication of hoping to break their own record time and again with a new facility that they began assembling in orbit in 1986.

This is the space station called Mir, meaning peace. According to American space experts, the Mir program is the clearest manifestation of the long-term Soviet goal of establishing permanent settlements off-planet — first on orbital stations, later on the moon and eventually on Mars.

The core of the Mir station is a long facility not much bigger than Salyut but more advanced in electronics, computers and other modern technologies. A key feature is its six ports. At least two ports are equipped for docking Soyuz ferry craft that bring up crews and supplies. The other four ports are attachment points for other modules for expanding the size and capabilities of the overall station. In 1987, the first of these extra modules linked up with Mir, giving the crew more living space and an array of astrophysics instruments for exploring the heavens. (Following a practice expected to be more and more common, Soviet scientists are collaborating with Western European scientists in outfitting the experiment modules.) Similar modules for other scientific projects and materials-processing experiments will likely be next. Soviet officials predict they will have crews of ten to twelve cosmonauts occupying expanded space stations in the 1990s.

Two apparent deficiencies in space technology must be overcome before the Soviet program can complete its transition from the experimental Salyut flights and early Mir operations to an expanded space station capable of being operated on a permanent basis. The Russians must improve their rocketry and their spacecraft for ferrying crews. In short, they need a more powerful booster for lifting heavier payloads, and they need a reusable craft along the lines of the American shuttle. American intelligence sources indicate that progress is being made on both these fronts.

According to recent American satellite photographs of Soviet facilities, the Soviet Union is developing a booster rocket for a version of the space shuttle and a new family of big rockets similar to those used by the United States for the Apollo launchings. Some reports estimate that the biggest of the new Soviet rockets could lift payloads of up to 150 tons into low orbits around Earth, about seven times more than the largest operational Soviet booster and five times more than the biggest American booster, the shuttle. (The Saturn 5, which launched the Apollos, is no longer in use.) Such large boosters would presumably be the mainstay for launching the major components for the permanent space station the Russians have long talked about. American sources believe that the Soviets are planning a space shuttle that "differs from the U.S. shuttle only in the respect that the main engines are not on the orbiter." This presumably means that the Soviet shuttle would not be as fully reusable as the American craft.

When will these two new elements of the Soviet program be ready? Probably soon. However, the Soviet experience in new technologies has been sluggish and troubled. Earlier attempts to develop a Saturn 5-class booster in the 1960s and 1970s ended in explosive disaster on the launching pad. Until such new technologies are proven ready, the Soviet manned space program is not likely to move forward in more than small incremental steps.

Even so, Soviet cosmonauts will probably be living and working in commodious orbital stations before American astronauts. Although Americans gained some experience in long-duration flight during the Skylab Project in 1973–1974, no attempt was made to follow through with a more advanced space station development until 1983. The Reagan Administration established as the next goal the deployment of a permanently manned, multipurpose space station by the mid-1990s.

Exactly what the station will look like and how it will be used have not been decided, pending further conceptual studies by engineers working for contractors for NASA. But James M. Beggs, the NASA administrator who finally won presidential approval of the project, has given the following outline of the station's possible functions. Astronauts and visiting scientists, as many as six at a time, could spend three to six months using the station as:

¶ A laboratory in space, for the conduct of science and the development of new technologies.

¶ A permanent observatory, to look down upon Earth and out at the universe.

¶ A transportation node where payloads and vehicles are stationed, processed and propelled to their destinations.

¶ A servicing facility where these payloads and vehicles are maintained and, if necessary, repaired.

¶ An assembly facility where, thanks to ample time on orbit and the presence of appropriate equipment, large structures are put together and checked out.

¶ A manufacturing facility where human intelligence and the servicing capability of the station combine to enhance commercial opportunities in space.

¶ A storage depot where payloads and parts are kept in orbit for subsequent deployment.

"Perhaps more important than any of the individual points, however," Mr. Beggs said, "is my belief that a space station could also lead to important activities and functions that we presently cannot even predict today. Were NASA to have a station, it could represent a fundamentally new and versatile capability to support activities in space over the next thirty years."

Although the details have not been worked out — and final approval may be delayed by further opposition by some political and scientific forces — the early phase of the program, beginning in about 1994, would see the launching by the space shuttle of several modules that would be fitted together to form the core of the station in an orbit several hundred miles above Earth. The core would contain the living quarters for the crew, the control center, data-processing units, solar-power arrays and docking ports for the shuttles. Out from the core, over the years, would be added clusters of laboratories, manufacturing facilities and service stations for the shuttles and orbital tugs. These little ships, either manned or unmanned, would move from the space station out to satellites in higher orbits to supply them or retrieve them for maintenance back at the station. More advanced tugs might one day ferry back and forth to lunar bases.

The space station, thus, is conceived as a system that could evolve and expand as the needs arise. No one can predict when the space stations, either American or Soviet, will begin to take on the sleek, comprehensive aspect of their science-fiction forerunners, such as in the movie *2001: A Space Odyssey.*

By the turn of the century, astronauts, scientists, technicians and perhaps the occasional citizen-observer will be occupying the

American space station through the years without interruption. This could become so much a part of "ordinary" life that their comings and goings will be without fanfare.

Under arrangements now being negotiated, people from many countries, in Europe, Asia and the Americas, will be working at the facility and sharing in the construction and operation. The European Space Agency is expected to provide a $2 billion laboratory module to be attached to the complex. Likewise, Japan is planning a $1 billion module, and Canada has promised to provide a "garage" for servicing satellites. Relations with these potential partners have been strained, however, by their fears that the United States would permit them little voice in station management and would insist on turning much of the facility over to military operations.

Someday, in a better world, the Soviet and American space stations might be operating in cooperation rather than competition.

More in the spirit of Pioneer 10, though, will be the exploits of the unmanned ships of exploration that will be going forth in the years ahead. They promise new insights into the other worlds of the solar system and a penetrating look toward the edge of the universe and the beginning of time.

One spacecraft, Voyager 2, is already en route to the more distant planets. After exploring Jupiter and Saturn, the craft made its way to a rendezvous with Uranus in January 1986. If its health continues, Voyager will pass close to Neptune in 1989. Then it will be recorded that man in the twentieth century visited for the first time, through its automated surrogates, all the planets of the solar system save Pluto.

Until recently, interplanetary exploration was largely an American endeavor. The Soviet Union has had considerable success sending unmanned craft to Venus, but no success at Mars, and has never ventured out to the outer planets. But interplanetary space will become more international, a change heralded by four spacecraft that greeted the return of the celebrated Halley's comet in 1986.

The Soviet Union sent two Vega spacecraft to fly by Venus, dropping off a couple of landers there, and then headed for encounters within 6,000 miles of the comet in March 1986. The craft carried cameras and instruments for investigations of the comet's solid nucleus as well as its surrounding cloud of dust and particles, the coma.

The European Space Agency, a consortium of thirteen Western European nations, dispatched a single craft, named Giotto, for an encounter within 600 miles of Halley's nucleus, also in March 1986. Soviet scientists shared Vega findings immediately with the Europeans so that they could make necessary last-minute changes in Giotto's trajectory and with ground observers so that they could adjust their instruments. Giotto also carried a camera and instruments for investigating the comet's dust, gases, charged particles, chemical processes and magnetic properties.

In August 1985, Japan launched its first interplanetary spacecraft for a flyby of Halley's comet in March 1986. The small craft, developed by the Institute of Space and Astronautical Science, carried an ultraviolet camera and a charged-particle analyzer for observations as it passed within about 60,000 miles of the comet.

Conspicuous by its absence from the greeting party was the United States. Proposals for an American Halley mission were rejected because of lean space budgets made leaner by the expense of the shuttle development.

Flushed by their success with the Halley missions, and growing confident enough to speak more openly of their goals far in advance, Soviet scientists are pushing ambitious plans to explore Mars in the 1990s. They expect to send craft to orbit Mars and probe the surface of the tiny Martian moon Phobos. Other craft are to follow: one to orbit Mars and send a penetrating instrument to study Martian geology, and another, perhaps in 1996, to pick up soil samples for return to Earth.

Momentum is building within NASA and the American scientific community for embarking on a major program of Mars exploration. A Mars observer craft, designed to orbit the planet and study its geology and climate, will be launched in the early 1990s. By the end of the century, if approval is gained in time, an American craft could be landing on the red plains of Mars bearing a roving vehicle. The robotic vehicle could be sent rambling over the planet, gathering data and collecting soil samples. The soil would then be transferred to a capsule for a flight back to Earth.

The sample-return mission, says William Quaid, an official of NASA's solar system exploration division, "appeals to NASA as something for science which is also a technical challenge to engineers at NASA."

All these efforts are considered the necessary preliminary steps to any attempt to send people to Mars, perhaps in the second or third decade of the twenty-first century. A space station could be the staging site for the assault on Mars. Although neither the United States nor the Soviet Union has an approved program for such an undertaking, the goal is never far from the sight of many space planners in both countries.

If humans are not likely to venture much farther out into space for a long, long time, they will be getting striking glimpses of incredible distances. This is the promise of the Hubble space telescope, which was to have been launched from a space shuttle in 1986 and is now expected to be deployed sometime in 1989, if the shuttles are back in service. Astronomers can hardly wait for the ninety-four-inch optical telescope.

The complete telescope assembly, a log-shaped tube forty-three feet long and fourteen feet wide, weighing twelve tons, is one of the first major space facilities designed to take full advantage of the space shuttle's flight capabilities. Not only will it be deployed by the shuttle, but it will be visited periodically by shuttle crews. They should be able to replace failed components or to install new instruments. From time to time, as the telescope drifts to lower altitudes, the shuttle will give it a boost back up to its regular 310-mile-high orbit, thus assuring the telescope's steady operation for several decades.

Although less than half the diameter of the 200-inch telescope at Mount Palomar, California, the space telescope should bring into sharp focus stars and galaxies never before seen because of its vantage point far above the earth's obscuring and distorting atmosphere. From there, stars would not seem to twinkle, the most obvious effect of atmospheric interference. From there, stellar radiations in the ultraviolet or infrared wavelengths, which are absorbed by the earth's atmosphere, would be observable.

Consequently, the space telescope should enable astronomers to study stars fifty times fainter than the dimmest celestial object yet detected and to encompass some 350 times more volume of space. Images transmitted by the telescope's instruments could reveal radiations that emanated 14 billion years ago. Scientists are not sure of the universe's age, but their various hypotheses lead to estimates of from 10 billion to 20 billion years — in which case the telescope should see objects at the edge or almost at the edge of the universe, if any objects are there to be seen.

According to John N. Bahcall, of the Institute for Advanced Study at Princeton, scientists working over the years with the telescope will be addressing such cosmic questions as: How big is the universe? How old is it? Did it have a beginning? Will it have an end? Will observations of black holes or quasars reveal new laws of physics? How are stars formed? What kinds of undiscovered things are there in outer space?

One of the telescope's earliest discoveries, astronomers suspect, could be the first known planetary system of another star. Astronomers have long assumed that the sun is not the only star with planets and that, if there is life elsewhere in the universe, it will inhabit some of the planets of some faraway stars. There have been tantalizing hints, through infrared astronomy, of dense material orbiting at least two major stars, suggesting perhaps planetary systems in the making. To find unequivocal evidence of other planets far beyond the solar system, worlds perhaps like our own, would be one of the most sensational discoveries of all time, both scientifically and philosophically.

At about the same time the space telescope is launched, a spacecraft named Galileo will embark for Jupiter, the planet first explored by Pioneer 10. The craft will fire an instrumented probe into the dense Jovian atmosphere to assess its physical and chemical properties. Then the craft will orbit Jupiter and thread its way among the four major Jovian satellites for extended observations.

Other planetary expeditions in the coming decades will likely include an orbiter to map Venus with high-resolution imaging radar, a probe to the Saturn moon Titan to determine the chemical composition of its atmosphere and the nature of its unseen surface, flybys of asteroids and a rendezvous with and extended flight alongside a comet.

THE COLONIZATION OF SPACE

Sometime in the twenty-first century, the colonization of space will begin. In its examination of the next fifty years in space, the President's National Commission on Space concluded in 1986: "The future will see growing numbers of people working at Earth orbital, lunar, and, eventually, Martian bases, initiating the settlement of vast reaches of the inner solar system."

Even if some of the projections are highly speculative, the prospects are exciting and, most planners say, well within the reach of modern technology. Using Earth-orbiting space stations as stepping stones, astronauts will land on the moon again and erect camps for permanent occupation. These lunar bases would include a research laboratory for science, materials processing and surface operations. To be self-sufficient, the base would carry out agricultural experiments using lunar soil and recycling water, oxygen and carbon dioxide. Exploitation of moon materials will be the major activity. Oxygen extracted from lunar minerals would be used not only to support the lunar base but to supply spaceships plying Earth-moon routes. Lunar soil would be mined for iron and titanium. Magnetic rakes could concentrate small quantities of metal from meteoritic material in the soil. Greater amounts of iron and titanium could be obtained from ilmenite through processes run by solar energy. These materials could be used to build large space structures, either on the moon or in Earth orbit. Since the moon's gravity is weak, it might become less expensive to ship oxygen and materials from the moon to space stations than from the earth.

When will we see permanent bases on the moon? In the next twenty-five years, perhaps. American scientists and engineers have already begun drawing up conceptual plans for such operations, and American officials have begun talking about lunar bases as a promising goal for the twenty-first century. One conclusion of the National Commission on Space was: "The proximity of the Moon, its inherent scientific interest, and its potential value for resources to be used in space all lead us to recommend: Establishing the first lunar outpost within the next 20 years, and progressing to permanently occupied lunar bases within the following decade."

Much of the talk at a 1984 meeting sponsored by NASA and entitled "Lunar Bases and Space Activities in the 21st Century" had the ring of science fiction. But the engineers insisted that they were simply extrapolating on current technologies when they spoke of the new propulsion systems, robotics and closed-cycle environmental systems necessary for human colonies on the moon or Mars. Whether their projections are realistic depends largely on the economic health of the world powers in the twenty-first century and a commitment by the affluent nations to direct their space efforts toward space stations and lunar and Martian bases.

Thomas O. Paine, a former NASA administrator and the head of the National Commission on Space, addressed the economics of such bold endeavors. He estimated that it would require about twice NASA's $12 billion annual expenditures during the Apollo moon-landing program to develop a colonization drive. In a report to the meeting, Dr. Paine said: "The task is more challenging, but greatly advanced space technology will provide far more capability per dollar. If a comparable sum is spent by the Soviet Union and other technically-oriented nations, as appears likely, terrestrial investment in developing Luna and Mars would average around $50 billion (1985 dollars) per year. To visualize this level of effort remember that it's about the annual output of the Ford Motor Company, and substantially below America's rocketing $60 billion annual expenditures on illegal drugs. Over a century this would represent a cumulative total investment to occupy and develop new worlds of about $5 trillion (1985 dollars). I believe that this is a reasonable price tag. If the annual Gross World Product in this period rises from today's $12 trillion to the projected $100 trillion in 2075, a $5 trillion investment in Mars would represent about a tenth of a percent of the 21st century's cumulative GWP of $5,000 trillion. A prospering Earth will easily be able to afford to initiate the evolution of terrestrial life on two more worlds."

The scenario for such extraterrestrial exploits, according to Dr. Paine and other space planners, might go something like this:

In the 1990s, both the United States and the Soviet Union will establish large stations in low-Earth orbit. These stations will grow in size and function over the years. At first, they will serve primarily as research and manufacturing facilities. Later, they will become bases for assembling spacecraft for use as ferries to the moon and out to more distant orbits, for refueling and servicing spaceships, and for mounting manned flights to Mars. Soviet officials have predicted that they will send men to Mars, though they never fix a date to their predictions. The first American-manned Mars expedition might shove off from one of these orbiting spaceports by the year 2010, though such projections are probably overly optimistic. This will also be the time when Americans will be returning to the lunar surface to establish permanent research bases for testing closed ecology communities and robotic systems for lunar mining and manufacturing.

Between 2015 and 2025, the first lunar community supported by robotic mining and closed-cycle agriculture will start experimental operation, powered by automated solar-energy stations. Oxygen will be separated from lunar rocks to refuel surface-roving vehicles and surface-to-orbit shuttles. One-way freighters with nuclear generators will begin hauling heavy structural materials to Mars to set up bases on that planet.

In another ten years or so, the first children will be born in the thriving lunar colony, now numbering several thousand people. People will have pioneered the Mars base by then. By 2045 the first Martian baby will have been born. One hundred years from now, according to these projections, the populations of the lunar and Martian settlements will exceed 100,000 each and have close to a million equivalent robots. They will be largely self-sufficient and thriving on a lively trade with the people back on Earth. Mankind thus is evolving independently on two new low-gravity, resource-rich worlds.

Whether all this is fantasy or future reality is, of course, hard to say. Who would have predicted a century ago that human beings in the twentieth century would walk on the moon and send probes on trajectories leading beyond the solar system?

Wernher von Braun, the German-born rocket engineer who was instrumental in mankind's pioneering thrust into space, once spoke of the problems of forecasting the future at the same time he expressed confidence that the future would be literally beyond our imagination. He wrote: "Only a miraculous insight could have enabled the scientists of the eighteenth century to foresee the birth of electrical engineering in the nineteenth. It would have required a revelation of equal inspiration to foresee the nuclear power plants of the twentieth.

"No doubt, the twenty-first century will hold equal surprises, and more of them. But not everything will be a surprise. It seems certain that the twenty-first century will be a century of scientific and commercial activities in outer space, of manned interplanetary flight and the beginnings of the establishment of permanent human footholds outside the mother planet Earth."

Such endeavors may also represent mankind's last chance to heal some earthly divisions through cooperation in space. A lunar base, NASA officials have said, would be an enormous challenge requiring new technologies and large economic resources, all of which implied, they added, the need for "international sharing of risks and benefits."

As a beginning, the United States expects to have Japan and several European nations join in the development of its space station. The European Space Agency is developing an increasing sophistication in space technology. It gained invaluable experience in manned flight operations through its development of Spacelab, a unit flown in the cargo bay of the space shuttle. Its Ariane rockets will become increasingly reliable haulers of payloads to orbit. Future Arianes could even be used to launch reusable shuttlelike vehicles; on the drawing boards are plans for a European shuttle called Hermes, which could be operating by the end of the century. Japan's electronics and robotics technology will also be a great asset in any future space settlements.

A lunar base might lead to cooperation among many nations, following the pattern of some of the international scientific activities in Antarctica. "An internationally developed lunar base," James Beggs, the former NASA chief, said, "might even prove an irresistible lure to the Soviets." Predicting political

behavior, however, can be the most hazardous aspect of futuro-
logy. The trend now seems to be running toward extending earthly
rivalries into space, including the weapons of war being fashioned
by these rivals.

But it does seem, barring the catastrophe of nuclear war, that
human beings will be spreading out into the solar system in the
next century, exploring and occupying a new frontier. Compared
with our lives in the late twentieth century, their lives may seem
incredible, as our lives would seem to anyone from the late
nineteenth century.

We can be sure that the space explorers and colonists decades
from now will look back on our spacefaring as a brave beginning,
but only a beginning. To them our spaceships will seem unbeliev-
ably primitive, like the caravels of Columbus. For by then, the
engines powering their wide-ranging flights will run on nuclear
energy, sunlight and perhaps sources of energy we have yet to
conceive.

Some of these propulsion systems have already undergone
preliminary testing. Ion engines, for instance, run on sunlight
collected by solar panels and converted to electricity, which is
used to charge electrically, or ionize, mercury vapor. The charged
mercury gas is concentrated and expelled in a steady, flameless,
violet exhaust — ion drive. Since a small tank of mercury would
be sufficient for years of flight, a spacecraft operating on ion drive
could be sent on long journeys to comets and asteroids or shuttle
out to supply distant satellites.

Someday, ships may be scudding through interplanetary
space like the caravels of old. They will be sailing in a "breeze"
caused by the light of the sun, sailing to sunbeams. Photons of
sunlight have no mass, but they do have momentum. They exert a
pervasive pressure that can push against gossamer sails in the
vacuum of space. Square sails like huge kites or long crisscrossed
sails could be spread out hundreds of feet, or miles perhaps, to
deflect the light as if it were wind. Preliminary work on solar
sailing is being done in Czechoslovakia, France and the United
States.

Solar-sailing enthusiasts envisage regattas in which unmanned craft race to the moon, which could be one of the sports of the twenty-first century.

They also believe that a mission to return a soil sample from Mercury might be more easily accomplished with sails than with rockets. Another sailing might visit a number of asteroids, in the manner of Captain Cook going from island to island in the Pacific. Or a ship might sail the heavy freight to Mars, getting there in time for use by the first astronauts, who would arrive by faster, smaller rocket-powered craft. Someday, perhaps, if dreams take flight on sunbeams.

Chapter Eight

Grand Unification Theories:
Faith in Ultimate Simplicity

By Timothy Ferris

Timothy Ferris is the author of The Red Limit, Galaxies, SpaceShots, *and other books on astronomy and astrophysics. He is professor of journalism at the University of California at Berkeley.*

> Great things are made of little things.
> — Robert Browning

Since the dawn of history, Man has pondered the riddle of the origin and structure of the universe. Pondering, however, didn't get Man very far. In the absence of hard facts, pure contemplation tended less to enlighten the mind about the universe than to turn it into a jack-o'-lantern, projecting old ideas against the sky rather than learning new ones from it. It was largely through unfettered contemplation that the ancient Sumerians, living as they did at a confluence of rivers, ascribed creation to a kind of mud-wrestling match among the gods, and that the early Christians in their enchantment with heaven assigned to the angels the job of pushing the planets along in their orbits across the sky. These ideas are not without a certain charm, but they tell us more about ourselves than about the outer world.

Science, a relatively recent development in the evolution of human thought, takes a less ambitious approach; scientists tend less to pronounce grand answers than to pose small, specific questions. As the Nobel prize–winning chemist François Jacob writes in his book *The Possible and the Actual,* "The beginning of modern science can be dated from the time when such general questions as 'How was the universe created?' 'What is matter made of?' 'What is the essence of life?' were replaced by more modest questions like 'How does a stone fall?' 'How does water flow in a tube?' 'What is the course of blood in the body?' "

This preoccupation with detail has exposed scientists to a certain amount of bemused belittlement ever since the days when Aristotle spent his honeymoon collecting marine biology specimens, but it gets results. "Curiously enough," Jacob writes, "while asking general questions led to very limited answers, asking limited questions turned out to provide more general answers." Charles Darwin's theory of the origin and evolution of life grew less from first principles than from Darwin's tireless scrutiny of the finches' beaks and horses' teeth that he crammed into the holds of the *Beagle.* Isaac Newton was induced to write the *Principia,* his greatest book, by considering the relatively narrow question — put him by Edmund Halley — of just what, exactly, would be the shape of the orbit of a comet if the forces of gravitation pulling on the comet were inversely proportional to the distance from the comet to the sun. Yet the results, in both cases, were genuinely universal: Newton's laws of gravitation describe the orbits of planets and stars and galaxies, and Darwin's evolution is the anchoring thread running through all earthly life.

Nowhere has the efficacy of studying the specific in order to learn about the general been demonstrated more dramatically than in recent years in particle physics, the science concerned with the smallest known structures in nature.

Ever since Leucippus and Democritus propounded the theory of the atom in the fifth century B.C., humanity has dreamed, however fitfully, of finding something fundamental — a variety of particle, perhaps, or a natural law — upon which all the wide world was built. The search quickened early in this century, when quantum mechanics began to discern order in the subatomic realm and accelerators were built with which experimental physicists could probe the structures of matter on scales smaller than the nucleus of an atom.

What was found at first seemed anything but simple. The atom, thought to be the fundamental particle, turned out instead to be composed of electrons orbiting a nucleus, and the nucleus, in turn, to be made of protons and neutrons. By the mid 1930s four elementary particles had been identified — protons, neutrons, electrons and neutrinos — and hopes ran high that this quadrumvirate might represent the foundations of reality. But those hopes were dashed when a bewildering variety of new particles turned up. The roster of allegedly "fundamental" particles soon climbed to more than 100: There were the bosons and fermions, the hadrons and leptons, the quarks and gluons and the charmed, strange mesons that made up what came to be called, not entirely kindly, the particle "zoo." As theories grew ever more elaborate in order to explain new experimental results, particle physics was denounced by skeptics as the modern equivalent of Ptolemy's Earth-centered cosmology, to which ever more complex hierarchies of epicycles had to be added in order to square its erroneous assumptions with the observed motions of the planets. Science seemed far indeed from glimpsing the elegant unity thought to underlie the wild diversity of nature.

But the situation has improved considerably in recent years. New, unified theories have appeared to point the way to a bedrock of simplicity beneath the myriad subatomic particles. The theoretical progress is ongoing, but real, and experimental testing of the new ideas challenging, but perhaps possible.

Most important, the new theories have a cosmological connection. Though they are concerned with the very smallest structures in the universe, they may well shed light on the largest question of them all, that of how the universe began.

The key to the confluence of large and small is that the new theories indicate that many of the complexities of nature — perhaps all of them — would disappear under conditions of extremely high energy. The universe is understood to have originated in just such a state — in the "big bang," the explosion that initiated the expansion of the universe. If the unified theories are correct, therefore, the simplicity long sought by science is not merely an abstraction or an aspiration, but a fact of cosmic history. The theories suggest, in other words, that the universe began in a state of simplicity, traces of which may be glimpsed today by replicating or envisioning how matter and energy behave under high-energy conditions like those that ruled the universe at the onset of time.

One way to understand the new unified theories is to approach them by way of the four fundamental forces of nature — gravity, electromagnetism and the "weak" and "strong" nuclear forces. Each force plays a distinct role in the workings of the world. Gravitation rules on the cosmological scale, binding the stars, planets and galaxies together. Electromagnetism holds atoms together as molecules, brings us light from the sun and stars and fires the synapses of the brain. The strong and weak nuclear forces, very short in range, keep protons and neutrons bundled together in the nuclei of atoms and govern the transmutation of unstable atoms via radioactive decay.

Reducing the workings of the universe to a system of only four forces is cause for celebration in physics, of course, but for frustration as well. The scheme is neat, but in some respects arbitrary. Why are there four forces, rather than a dozen or two or one? And why do the forces differ so profoundly in character? Gravitation is weak — a hundred million trillion trillion trillion times weaker than the strong nuclear force — but it affects all forms of matter and energy, while the strong force effects only one class of particles, the hadrons, and is ignored by another, the leptons. Electromagnetism manifests itself as both positive and negative electrical charge, but there appears to be no such thing as negative gravity. And so on; the list of unexplained differences among the forces is long.

The standard model, as it is called in contemporary physics, permits accurate predictions to be made for the outcome of a great variety of events, from the gravitational interaction of a pair of stars to the workings of the strong force in the thermonuclear reactions that power the stars. But each force must be accounted for by a different set of equations. In all, some twenty different parameters must be introduced into the equations; they work, but nobody is certain *why* they work, and this rankles just about any scientist worth his or her salt. Says Steven Weinberg, a theoretical physicist at the University of Texas at Austin, "Our job in physics is to see things simply, to understand a great many complicated phenomena in a unified way in terms of a few simple principles."

It was Weinberg who came up with a key insight in the search for simplicity. He was thinking not about large but about small things — specifically, about the minuscule particles that carry electromagnetism and the weak nuclear force.

Particle physicists envision matter and energy as composed of particles — hence the name of their discipline. The particles that convey the four forces differ considerably from one another, as do the forces themselves. The differences are particularly acute in the case of electromagnetism and the weak nuclear force. Electromagnetism is carried by photons, which have no rest mass at all, while the weak nuclear force is transmitted by weak bosons, which are relatively massive and ponderous. Nonetheless, one autumn morning in 1967, it occurred to Weinberg while he was driving to his office that despite the obvious differences between photons and weak bosons, an obscure but genuine kinship might link them.

Weinberg's hunch, along with his subsequent work and that of colleagues, did much to bridge the perceptual gulf separating the weak force from electromagnetism and gave rise to the first of the new unified theories, the "electroweak" theory of particle interactions. The electroweak theory demonstrates that electromagnetism and the weak nuclear force may be regarded as aspects of a single, electroweak force. For their contributions to developing the electroweak theory, Weinberg shared the 1979 Nobel prize in physics with Sheldon Glashow of Harvard University and Abdus Salam of the Imperial College of Science and Technology in London and the International Center for Theoretical Physics in Trieste.

The unification of forces envisioned by the electroweak theory — and by the more ambitious grand unified theories that sought to unify the strong nuclear force with the electroweak — would manifest itself only if the environment were hotter than can be attained in the highest-energy particle accelerators on

Earth, or, for that matter, anywhere else in the contemporary universe. It was largely for this reason that the theory seemed unrealistic and was at first ignored by the scientific community. If the energy that could forge the two forces into one was unavailable, what had the theory to do with the real world?

The answer came from the opposite end of the scale of size — from cosmology, the science of the very large, the study of the structure and dynamics of the universe as a whole.

The universe is expanding, the far-flung galaxies receding from one another at velocities directly proportional to the distances separating them from one another. Galaxies 32 million light-years apart are receding from each other at a velocity of some 500 kilometers per second, galaxies ten times farther apart are receding ten times faster, and so on. This literally universal phenomenon was predicted, implicitly, by Einstein's general theory of relativity, published in 1915, but for lack of observational evidence, the prediction was at first discounted by almost everyone, including Einstein himself. Then, in one of the most astonishing coincidences in the history of science, evidence of the expansion of the universe soon was discovered, in the late 1920s, by American astonomers who knew nothing of Einstein's theory.

Another way to say that the universe is expanding is to say that it is thinning out — that the amount of matter in a given volume of space is, on the average, steadily decreasing. If so, the universe must, long ago, have been in a state of much higher density. Theoretical investigations of this prospect were first carried out by the Belgian cosmologist Georges Lemaître and the Russian-American physicist George Gamow, and have since been elaborated upon by other scientists. Their research indicates that the expanding universe began as a dense, hot fireball — the big bang, as the British astronomer Sir Fred Hoyle waggishly dubbed it.

Gamow and his colleagues Ralph Alper and Robert Herman noted that if the universe once was hot, some of the heat from the fireball should still be around, permeating the universe. Thinned out by the subsequent expansion of the universe, the ubiquitous heat should today amount to only a few degrees above absolute zero. At the time there was no way to test this prediction. But in

1965, the American radio astronomers Robert Wilson and Arno Penzias accidentally detected just such residual heat, using a state-of-the-art radio telescope. Known today as the cosmic background radiation, the dying murmur of the big bang has been widely studied and proves to conform to the characteristics predicted by the big bang theory. Thanks to these and other studies, the reality of the big bang stands as one of the best-established tenets of cosmology today.

Here, then, was a time when the energy levels envisioned by the unified theories would have existed. During the first moments of the big bang, temperatures throughout the universe would have been so high that unification could have pertained, and the four forces functioned as but three forces or two or perhaps only a single primordial force. The big bang lends a historical dimension to the physicists' quest for a unified theory: it implies that the universe began in a state of simplicity that has since been buried beneath the detritus of 20 billion years of cosmic history.

Interesting if true, as Dr. Johnson used to say. But are the theories true? A direct test would require recreating the conditions that existed in the big bang, and that is what particle accelerators do. The energy they generate — albeit for but an instant, and on a very small scale — replicates the environment that once embroiled the entire universe. The more powerful the particle accelerator, the earlier the epoch in the big bang that it can recreate. Although no accelerator has yet been built powerful enough to replicate the conditions in which the weak and electromagnetic forces are thought to have functioned as one force, it was just possible to bring an existing accelerator to a power sufficient to test some of the indirect consequences of the theory.

The electroweak theory had predicted the existence of a previously undetected variety of particles called neutral intermediate vector bosons. Like the salamanders of mythology, intermediate vector bosons can thrive only in fire; the amount of energy needed to flush them out was more than could be attained by any accelerator in existence at the time that the theory first was propounded. But by 1983, the giant particle accelerator at CERN, near Geneva, had been souped up until it could achieve the energy required to test the contention of the electroweak theory that neutral intermediate vector bosons actually exist.

The driving force behind the CERN experiment was Carlo Rubbia, a rotund and convivial physicist who commutes between Harvard and Geneva so frequently that he is estimated by his collegues to have a lifetime average velocity, day and night, of forty miles per hour. Rubbia teamed up with Simon van der Meere, a resourceful engineer, to increase the energy of the CERN accelerator by equipping it so that particles of antimatter would collide with particles of ordinary matter.

The origin of this bold idea lay in a theory propounded twenty years earlier by Richard Feynman of the California Institute of Technology. Feynman had noted that particles of antimatter — which are identical to those of matter but have opposite electrical charge — could be regarded as ordinary matter moving in reverse time. Some scoffed at Feynman's theory, but none could fault his logic, and eventually physicists grew accustomed to entertaining the idea that antimatter particles wander through the universe in backward time, like actors in a film projected in reverse. The idea seemed amusing, albeit useless.

Rubbia and van der Meere proposed to put it to use. Their plan was to inject antimatter particles into the CERN ring, where magnetic pulses are employed to accelerate particles to nearly the velocity of light. The antimatter particles, sensing the pulses in reverse time, would orbit the ring in the opposite of the normal direction. If they could then be directed into the oncoming stream of particles of ordinary matter, the result would be a headlong collision at extremely high velocity, resulting in a tiny explosion of unprecedented intensity.

Big science at its most daring, the CERN matter-antimatter collider experiment absorbed the efforts of 120 physicists from eleven universities, as well as whole platoons of engineers, and it encountered technical problems that took years to overcome. But by the summer of 1983, thousands of matter-antimatter collisions were taking place in the accelerator, and neutral intermediate vector bosons were detected. A key prediction of the Glashow-Weinberg-Salam theory had been upheld, a feat for which Rubbia shared the 1984 Nobel prize in physics with van der Meere.

Meanwhile, other ingenious experiments were being mounted to test the claims of the grand unified theories. The grand unified theories contained the startling implication that matter itself might be but a passing phase in the evolution of the universe. Specifically, they mandated that protons — particles found at the heart of every atom — do not last forever but must eventually decay. The proton decay rate predicted by the grand unified theories is so slow that enormous quantities of protons — at least ten thousand billion billion billion of them — must be monitored in order to have a 50 percent chance of seeing just one proton decay in the course of a year.

Vast collections of protons accordingly were amassed in the form of tanks of water or stacks of concrete and steel, and were monitored at test sites buried in the bowels of a salt mine in Ohio, in a gold mine in India, in a railroad tunnel beneath the Alps and in other subterranean sites. (The experiments are run deep underground in order to minimize contamination by high-energy particles streaming in from space, which can collide with particles in the experimental chamber and mimic the effects of proton decay.) By the late 1980s the results of these long-term proton watches were still inconclusive, though they had set a lower limit on the proton half-life that invalidated at least the simplest version of grand unified theory. Other versions of the theory predicted a longer protein half-life, and could not be tested until the experiments had continued to run for years longer.

Meanwhile, considerable excitement was being generated in the scientific community by the development of supersymmetry theories that might present a unified account of all four forces. The supersymmetry theories, called SUSYs for short, purport that the universe originally was built on more than the four dimensions — the three dimensions of space plus one for time — that characterize nature today. According to the SUSYs, the universe could originally have had, say, ten dimensions, of which all but four rapidly collapsed, at the onset of the big bang, into tiny objects too small to be noticed today. Conceivably, these objects might be the subatomic particles themselves — quanta of matter and energy that originated by condensing out of pure space.

Support for the supersymmetry hypothesis came from the

superstring theory, which portrays the subatomic particles, not as infinitesimal spheres, but as tiny strings. Physicists found it conceivable that all the varied properties of particles, including their mass, electrical charge and spin, might be explained in terms of possible configurations of the strings — e.g., as lines, loops and X's.

By the late 1980s the prospects for supersymmetry theories — including their cousins the superstring theories — continued to look bright. Since they are inherently geometrical, SUSYs naturally incorporate relativity, which portrays the force of gravitation as a manifestation of the geometrical curvature of space, into the account of quantum mechanics; indeed, writes John Schwarz of Caltech, one of the pioneers of supersymmetry, "it seems impossible to construct a mathematically consistent theory of strings *without* including gravity." And, where previous unified theories had to be "renormalized" — i.e., subjected to mathematical manipulations to eliminate terms that otherwise went uncontrollably to infinity — SUSYs introduce terms that automatically cancel each other out, eliminating the unwanted infinities.

A price paid for all this is complexity. Physics in a ten-dimensional universe is an intricate business, and by 1987 theorists were settling in for a prolonged period of hard work trying to make sense of it all. "The easy problems were solved in the first few months," sighed an MIT researcher working in the field. "Now we're left with the hard ones." Complexity in itself, however, is not necessarily a bad sign — Einstein's general relativity equations are more complicated than the Newtonian equations they superseded — and in some ways it may amount to a virtue. Supersymmetry implies, for instance, that there are whole families of as yet undiscovered subatomic particles; the number of previously unknown particles envisioned under the SUSYs ranges from enormous to infinite. This could be good news for astronomers who have long wondered why 90 percent of the matter in the universe, its presence betrayed by the dynamical behavior of galaxies, is invisible; this "missing mass" could turn out to consist of clouds of particles whose existence was unsuspected until SUSYs came along.

If experiment is to keep pace with rapidly developing new

theories like SUSY, new and more powerful accelerators will have to be built to recreate conditions of even earlier cosmic history. In 1987, President Reagan announced his support for such a machine, the superconducting supercollider, a particle accelerator that would be sixty miles in circumference. Its power would be more than twenty times that attained by Rubbia's team at CERN. Its significance, however, is likely to go far beyond confirming or denying supersymmetry, which in any case was in the late 1980s still in too preliminary a state for the nature of its experimental tests yet to be clear. The real function of such a giant accelerator would be to explore realms of high energy, on scales of fine resolution, about which little is yet known. The data generated could as easily prompt fresh theoretical insights as test old ones.

Individual theories may stand or fall, but the unification approach is likely to persist for aesthetic as well as rational reasons. Supersymmetry in particular depicts the universe as having begun in a state of sublime perfection, the contemplation of which elicits such enthusiasm in the minds of theoretical physicists that some have taken to calling the very early universe paradise lost. As Stephen Hawking of Cambridge University puts it, "The early universe was simpler, and it was a lot more appealing *because* it was a lot simpler."

The picture that elicits the affection of the theorists is of a universe that began in a state of absolute symmetry, in which there was neither place nor time nor varieties of particles and forces. A cosmic fall from grace came during the first fraction of a second of the expansion of the infant universe, when the symmetries of genesis fractured as the universe cooled. Viewed by these lights, the contemporary universe resembles a jumble of broken symmetries, like the heaps of potsherds that archaeologists must painstakingly piece together if they are to glimpse the beauty of the original pot. "The laws of physics are simpler to discern and understand when one goes to higher energies," says Michael Turner, resident cosmologist at Fermilab, "because at high energies symmetries are manifest, while at low energies they are hidden."

The fall from grace is seen as essential to existence as we know it. Absolute symmetry is beautiful, but it is also sterile. Every motion breaks the symmetry of space, by creating a "here" different from "there." Every event breaks the symmetry of time, by creating a difference between "before" and "after." Life thrives on imperfection and is full of examples of the cosmological principle that one has to break symmetries to get things done, as surely as one must break eggs to make an omelet. Serve the eggs at a circular table at which, say, six diners are seated: etiquette aside, it doesn't matter whether the dinner guests choose to use the butter plate on their right or their left side; but the choice has to be made, and once one dinner guest has chosen a plate, the symmetry has been broken and all the others must follow suit. A light bulb may be manufactured so that it screws in clockwise, as most do, or counterclockwise, as do the light bulbs used by the New York City subway system in order to discourage riders from unscrewing them and taking them home. But the symmetry must be broken, one way or another, if any threaded light bulbs are to be manufactured. Similarly, a universe that remained forever in a state of perfect symmetry could not have fragmented into the four forces, the subatomic particles and stars and planets and living beings. In a very real sense, we may owe our lives to the fact that the universe is less than perfect.

The development of the unified theories promises to write a new chapter in the old dichotomy between the perfect symmetries envisioned by the human mind and the asymmetries that permeate the material world. We can think, for instance, of a perfect sphere, but few if any perfect spheres are found in nature. (Stars and planets look spherical at a glance, but upon closer inspection prove to be lumpy, distended, and flattened at the poles.) Spiral galaxies come in pairs, but one member of the pair is usually much larger than the other; human beings have two hands, but one is usually stronger and more adroit than the other; antimatter is the mirror image of matter — antimatter particles are identical to those of matter, but have reversed charge — yet, for some reason, the universe is made almost exclusively of matter and contains little antimatter.

Plato wrote of the asymmetry that distinguishes the eternal human soul from the mortal body that houses it, and regarded the imperfect, material world as but a shadow of an ideal realm of perfectly symmetrical geometrical forms. In Plato's *Symposium,* Aristophanes identifies asymmetry with the fall from grace; human beings, he says, originally spherical in shape, were cut in two by Zeus as punishment for the sin of pride.

Weinberg, in accepting the Nobel prize, struck a Platonic theme. "In the seventh book of the *Republic,*" he recalled, "Plato describes prisoners who are chained in a cave and can see only shadows that things outside cast on the cave wall. When released from the cave at first their eyes hurt, and for a while they think that the shadows they saw in the cave are more real than the objects they now see. But eventually their vision clears, and they can understand how beautiful the real world is.

"We are in such a cave," Weinberg continued, "imprisoned by the limitations on the sorts of experiments we can do. In particular, we can study matter only at relatively low temperatures, where symmetries are likely to be spontaneously broken, so that nature does not appear very simple or unified. We have not been able to get out of this cave, but by looking long and hard at the shadows on the cave wall, we can at least make out the shapes of symmetries, which though broken, are exact principles governing all phenomena, expressions of the beauty of the world outside."

Contemporary physics resurrects not only Plato's dream of a symmetrical perfection underlying nature, but also his vision of learning as remembrance. Socrates, coaching the untutored slave boy in the precepts of geometry, claimed to have demonstrated that the "soul has been forever in a state of knowledge," and that all real learning, therefore, is recollection. Generations of philosophers have remained skeptical on this point, and understandably so; at the least, Socrates underestimated his effectiveness as a teacher. But, if traces of cosmic history are indeed etched within every scrap of matter, it is tempting to speculate about the extent to which cosmic history is woven through the human mind as well. Perhaps the dream of an ultimate unified theory is a form of cosmological nostalgia, and human science is a means by which the universe contemplates its past.

Chapter Nine

Beyond the Knowable:
The Ultimate Exploration

By Holcomb B. Noble

Holcomb B. Noble is the deputy director of science news of The New York Times.

The twenty-first century promises to explain the unexplainable. It offers humanity opportunities to learn what until now has been unlearnable, unknowable. It is not hard to imagine the twentieth century as a kind of arbitrarily chosen point from which the twenty-first century splays out in virtually all directions and all dimensions, building on the known, explaining the unknown. This is no image of the idle futurist. It is meant, on the contrary, to serve as a serious suggestion: in the next century, once grand unification is achieved, once machine intelligence matches human intelligence, once space exploration fulfills its promise, once communications systems meet theirs, the fundamental dimensions of space and time themselves may possibly be altered or become alterable. Given the nature of current scientific research and experimentation, it is not inconceivable, for instance, that the future may somehow develop a greater capacity to peer into "the dark backward and abysm of time," or that it will explain other vast areas of the present unknown.

It is, in a sense, this idea of the unknown that makes any future at once unsettling and exciting. Two bits of the present unknown that now nag persistently at the human consciousness and somehow seem more and more knowable are phenomena largely regarded in the past as quaint whimsy: extrasensory perception and extraterrestrial intelligence. The scientific mind is the mind that thrives on incontrovertible truth, that demands concrete proof. That's correct, isn't it? If you can see it, smell it, hear it, touch it, taste it, it must be there. Conversely, if you can't, it must not. Well, perhaps. But perhaps not. Perhaps, the truly scientific mind is the open mind, the one willing to explore even what cannot be perceived to be there. Who is to say that intensive exploration will not detect a world of intelligence and communication that goes far beyond the sensorily observable? Are we absolutely certain that we will not begin to make unexpected and remarkable discoveries through channels that extend into events that defy all rational explanation? The creative scientific mind does not rule it out. Einstein insisted to the last that a cause could be found for every effect. But a great many scientists have come to accept the concept of irrationality and randomness, and they believe it may become more dominant in decades to come.

There is no convincing evidence to prove the existence of extrasensory perception — the act of perceiving communication in ways other than through the senses. But a substantial number of respected scientists began, at least by the 1970s, to consider it as a possibility. Dr. Mahion Wagner, a psychologist at the State University of New York, Oswego, published a survey in 1979 based on questionnaires he sent to 2,100 college and university professors throughout the United States; 1,188 responded. Of the natural scientists who replied, 9 percent said they accepted extrasensory perception as an established fact and 45 percent called it a likely possibility. Private individuals and corporations have donated hundreds of thousands of dollars for research, at Princeton and elsewhere, for such work as the attempt to change temperature readings or move objects by mere will of the mind.

Arthur Koestler, who believed that parapsychology would introduce a new Copernican Revolution, left $750,000 in his will in
1983 for a chair in parapsychology at a British university; Oxford
and Cambridge both turned the bequest down on the grounds that
its very acceptance would challenge the notions of rationality
implicit in their scientific research. But the University of Edin-
burgh accepted it.

The 1984 spring meeting of the American Academy for the
Advancement of Science, one of the nation's most important
scientific gatherings, held a symposium called "The Edges of
Science." The meeting was held within a basic framework of
skepticism. The emphasis was on what remains unproven. But the
fact that the symposium was held at all was significant. Take the
question of intelligence in outer space and the controversy over
UFO's, unidentified flying objects. Arthur C. Clarke, the science
fiction writer, told the gathered scientists that there was a time
when he took UFO's seriously. "Now," he said, "UFO's need a
period of malign neglect." Every report of another landing, every
story, every anecdote turns out to be just that — another report,
another story, another anecdote. "If there really is a landing, it can
be documented in thirty minutes." Indeed, he asked whether the
whole UFO sighting question could not benefit from recent
laboratory evidence that people can hallucinate so well that
electrical activity in their brain agrees with their vision.

But another speaker, the respected J. Allen Hynek, made the
contention that there have been just too many unexplained UFO
reports to be lightly pooh-poohed. "Computer analyses of 400
UFO cases," he said, "are totally inconsistent with those of
everyday objects and phenomena." He said it was wrong to
suggest that "independent witnesses could be so grossly misled as
to imagine simultaneously, for example, that a meteor could stop,
hover, and then reverse direction." Space engineer James Oberg
countered with just that suggestion. He said that a series of
dramatic sightings in the Soviet Union and South America
happened to follow nighttime rocket launchings in Russia and
Argentina. The "objects" were vividly reported: people said they
stopped overhead, hovered, then turned and chased cars. What

was clearly going on, he said, was that human fear and imagination were interacting with an observable physical event to produce a series of wildly erroneous reports.

Other scientists as well believe that most studies of unidentified flying objects, along with research into extrasensory perception, are flawed. They argue that the work is often pretentious pseudoscience and invariably produces questionable results. They contend, too, that the field is dotted with charlatans and naive incompetents. Yet the fact that the American Association for the Advancement of Science has begun consideration of the subjects is one indication among several others that these claims on the edge of science are now being taken more seriously, that they must begin to be investigated with more rigorous scientific standards. The possibility that they exist is no longer so absurd as to permit the matter to drop. Careful study, the association was implicitly stating, is called for.

ESP AS MILITARY WEAPON

Nowhere is the possibility that life, energy, power, intelligence or communication may exist on some previously undiscovered level taken more seriously than it is in the halls and conference rooms of the military. The military establishments of the world's two superpowers — the Soviet Union and the United States — are actively pursuing it. According to a variety of reports, the Pentagon has spent millions of dollars on secret projects aimed at investigating extrasensory phenomena — to see whether the sheer power of the mind can be harnessed to perform various acts of espionage and war. Could, for example, highly developed mental powers be used to penetrate secret enemy files? Or, could the unaided mind be used to locate submarines or blow up guided missiles in midflight?

Although the Pentagon denies that it is spending money on psychic research, assertions to the contrary have come forth in important writings on the subject and in interviews with scientists and with former officials of the Pentagon. What becomes clear from these sources is that the superpowers are, indeed, actively engaged in attempting to master such arts as extrasensory perception (ESP), telepathy (which is generally defined as thought transfer), clairvoyance (seeing things that are not yet there) and psychokinesis (influencing the course of events or the behavior of people or things, simply by bringing mental energy to bear). This research is done in the name of national defense. "The Defense Department would be derelict in its duty if it didn't pay attention to the long shots," Dr. Marcello Truzzi of the Michigan-based Center for Scientific Anomalies told William J. Broad of *The New York Times* science department in an interview in January of 1984. But he hastened to add that the fact that the department is looking into the phenomena does not make them real.

Well, what are some of those experiments? Dr. Russell Targ, a physicist at SRI International, an independent research organization that is an offshoot of Stanford University, describes some of them in a book he wrote with Keith Harary called *The Mind Race.* He says SRI has worked under a multimillion-dollar United States Defense Department contract to study psychic behavior or events. Its principal experiments were attempts at what he called remote viewing. Skilled individuals were said to be able to describe objects located thousands of miles away.

A viewer in California was asked to visualize a specific site in New York City that he was totally unfamiliar with and to type impressions into a computer of what he saw. Dr. Targ says those recorded impressions were "of a cement depression — as if a dry fountain — with a cement post in the middle or inside. There seemed to be pigeons off to the right flying around the surface of the depression." The test site, Dr. Targ writes, was the central fountain in Washington Square Park. It was, indeed, dry, had a post in the middle, from which water could be sprayed, and it was surrounded by pigeons.

According to Ronald M. McRae, who is the author of *Mind Wars* and is a former reporter for the syndicated columnist Jack Anderson, the Pentagon has spent about $6 million dollars a year on psychic research for the past several years. He contends further that such research encompasses projects undertaken over the course of the past thirty years by the CIA, the Army, Navy, Air Force, Marine Corps, NASA and the Defense Intelligence Agency. Indeed, former U.S. government officials confirm that the explorations into parapsychology are going on, although in most cases they will not identify individual projects. One former White House official in the Reagan Administration, however, did confirm in an interview with William Broad that psychic researchers were used by American defense planners to try to locate hidden MX missiles.

At one time the Pentagon had developed a $40 billion plan to hide the giant intercontinental MX missiles in a series of selected concrete bunkers. The missiles would be secretly transported, either by rail or by huge flatbed trucks, from one bunker to the next, hidden so that Soviet military planners would never know where to strike. Ronald McRae reports that the Pentagon set up experiments in which trained observers, using only the powers of their own minds, guessed the positions of the missiles with sufficient accuracy to raise doubts about the security of the hide-the-missile plan. Barbara Honegger, the former White House official, did not know whether the experiments were a factor in the ultimate decision to abandon the whole idea.

The Soviet Union is said to be striving to harness the power of telepathic communication and telekenetics. Martin Ebon writes in his book *Psychic Warfare* that the major impetus for the Soviet drive into psychic exploration also came from the military and the KGB. He contends that the Soviet Union was goaded into action by false reports that the U.S. Navy tried to communicate with the first nuclear submarine, the *Nautilus,* by thought transfer, as it cruised under the Arctic ice cap.

It's perfectly true that all this may be just so much hokum. Is it not possible, for example, as indeed many scientists forcefully argue, that both the United States and the Soviet Union are once again whistling in the wind; that, out of fear and lack of respect for each other, they are off on another arms race — except that this time the coveted weaponry may be nothing more than figments of their collective imagination? Indeed, some suggest that the real goal is disinformation. To give the enemy the impression that mind weapons were a reality would be to set him off on expensive but harmless diversions. Even many of the skeptics agree, however, that the military has no choice but to pursue the paranormal.

Some serious nonmilitary researchers have been at work on the parapsychological for years. Dr. Brian Josephson, a British scientist who won the Nobel prize for physics in 1973, has spent much of his time since then studying the paranormal, of whose existence he says he is 99 percent convinced. Dr. Josephson's work in physics led him to the discovery that electrical conductivity in an ultra-cold environment can be switched on or off with a magnetic field, a discovery now known as the Josephson effect.

But are the rigors of hard science or quantum mechanics really compatible with investigations of the paranormal and what the intelligent skeptic always regarded as the quackery of, say, the old professional mind reader? "You ask whether parapsychology lies within the bounds of physical law," Dr. Josephson said to an interviewer. "My feeling is that to some extent it does, but physical law itself may have to be redefined. It may be that some effects in parapsychology are ordered-state effects of a kind not yet encompassed by physical theory."

In fact, the discovery of quantum mechanics is itself preliminary evidence in the minds of some scientists that extrasensory phenomena do exist, or are at least possible. The quantum theory conveys the notion that subatomic particles may be able to communicate with each other instantaneously. The discoveries of Paul A. M. Dirac and others suggest to some that widely separated electrons talk to each other, acting in concert and telling each other what to do. These particles do this in apparent violation of a basic tenet of physics: no signal can travel faster than the speed of light. Could it really be true, as it seems to be inside an atom, that an event can happen so fast that it causes other things to take place before the event itself has even occurred? Scientists at the Applied Physics Laboratory of Johns Hopkins University have tried to prove there must be some mistake. This cannot be right, they contend: nothing happens before it happens. Nothing travels faster than the speed of light.

The French theorist Bernard d'Espagnat concedes that the theory is strange, all right. The notion of signals outracing light would lead, he says, to "bizarre paradoxes of causality in which observers in some frames of reference find that one event is 'caused' by another that has not yet happened."

But he points out in an article in *Scientific American* that the bizarre does indeed seem real. Tests, he said, show that, in some atomic processes, particles can be ejected in opposite directions and those particles appear to have no properties at all until they are measured. Then the very instant they are measured, they match one another in location and spin even though they are far apart. Their behavior is baffling unless there has been some kind of communication between them, some kind of data being transmitted faster than the possible — faster, in other words, than the speed of light, faster than at the rate of 186,000 miles per second.

Other tests of this hypothesis have involved the ejection of two photons, particles of light, shot out in opposite directions from an atom primed by energy injection, as from a laser. When the photons are observed, each is polarized, or oscillating, in the same manner. This subatomic ballet of precision is observed even though some physicists believe the photons were never synchronized until the very moment of observation. The rational scientist has always been extremely frustrated by quantum theory. He wants to believe in cause and effect, and yet when he looks at the oscillating photon there seems to be no such relationship. Einstein and some others, however, never accepted such randomness. Einstein recognized the evidence for quantum mechanics as valid, but he said, "An inner voice tells me it's not the real thing." He did not believe, as he put it, "that God plays at dice."

We do known now from the great advances in physical theory and in the developing technology of nuclear energy and solid-state electronics that quantum mechanics is itself real enough. True, we have not proved whether God is or is not playing at dice. True, there is no universal agreement among physicists that the bizarre behavior of subatomic particles is in fact the result of instantaneous particle-to-particle communication and not caused by some other as yet unexplained phenomenon. But the idea is alive and hotly debated, and it keeps the scientific mind wondering about its implications. The mind is set to thinking about ESP or thought transfer, for example. Couldn't that phenomenon be explained in some manner by quantum mechanics?

Is it not possible that data can be instantaneously transmitted from the brain of one person to another in much the same manner as the communication between subatomic particles? Might not ESP work similarly to those talking electrons? Might not as yet undiscovered thought particles be transferred in waves somehow analogous to those of photons? Quantum mechanics has seemed to prove again what science has illustrated over and over through the centuries: the lack of understanding of the why of an event or the misunderstanding of it does not negate its existence. Is it correct to conclude that, because we do not understand ESP, it isn't there? We don't understand much about quantum mechanics, yet we accept the notion that its discovery is one of the most important in the history of science. "No one understands quantum mechanics," says Nobel laureate Richard P. Feynman. Its effects are "impossible, absolutely impossible" to explain based on human experience. It may be equally true of ESP. It may exist. It may be important to human and physical behavior. Yet it may not be explainable until long after its discovery, if then. We still cannot explain gravity.

Musings about the unknown are by definition speculative, and these may be highly so. But if d'Espagnat is right about electron events happening before they happen, why couldn't thought be transferred before it occurred?

SEARCHING FOR LIFE AND ENERGY IN SPACE

Another unknown, equally imponderable, is whether there is intelligent life on other planets. Again, despite serious scientific research that has stretched out over more than a quarter of the twentieth century in time and billions of miles in space, no clear evidence has been discovered that life does exist beyond the earth. Yet the search goes on, financed in part by a $1.5 million annual budget from the U.S. space agency. If anything, it has become more intense than when it first began in earnest in 1959, the year Dr. Philip Morrison of MIT and his colleague Dr. Giuseppi Cocconi made the suggestion that other civilizations might be trying to communicate with one another on a specific radio wavelength. They speculated that such a wavelength might be at twenty-one centimeters on the electromagnetic spectrum, based on the radiation given off by free atoms of hydrogen, the most common element in the universe. At least forty-five astronomical search projects have been initiated in attempts to pick up such extraterrestrial communication, and some of them, such as those at Harvard and Ohio State universities, continue to focus on the twenty-one-centimeter wavelength. By the 1990s, a satellite called Cosmic Background Explorer should begin monitoring cosmic radiation and should assist in the search for radio transmissions at the twenty-one-centimeter wavelength.

During the past century, many scientists believed there might be life on Mars, and they proposed various methods of trying to send out signals from the earth. One plan, devised by the mathematician Karl Friedrich Gauss, was to plant a giant forest in Siberia in the shape of a right triangle. Another suggestion was made that squares be planted on each side of the triangle to illustrate the Pythagorean theorem. Or, canals might be dug in the Sahara in the form of a geometric figure, these scientists suggested, and then, with kerosene covering the water, ignited at night. This would become a literal signal of intelligence flashing across space brightly enough so that intelligent beings out there might see it and respond.

Although there is now no evidence to suggest that life exists anywhere as close as Mars, the idea of lighting up the world with a sign of intelligence to shine into space was revived in the fall of 1983. SETI-France, a group searching for extraterrestrial intelligence, planned to set out floodlights and illuminate the Greenwich meridian along a 160-mile strip from Villers on the English Channel to Trois, a town southwest of Tours. At the La Flèche Airport, which is precisely on the meridian, fifty flares would be illuminated in the form of a cross, and at Paris, 200 torches would be lighted in the Place du Pantheon. All these lighting events would coincide with the almost certain appearance in space of intelligent human beings: that is to say, with the scheduled launching of one of the American space shuttles, Columbia. The crew of Columbia flying in space in orbit around the earth would see visible signs of people there trying to communicate with them. Bad weather prevented the SETI plan from being carried out, as impracticality perhaps had interfered with Gauss's idea for the

Siberian triangular forest. But they both symbolized the profound belief among thoughtful people that we are not alone.

There have been tantalizing hints, through infrared astronomy, of dense material orbiting at least two major stars, suggesting perhaps planetary systems in the making. To find unequivocal evidence of other planets far beyond the solar system, worlds perhaps like our own, would be one of the most sensational discoveries of all time, both scientifically and philosophically. In the search for these worlds in the 1980s, an entirely new era, the Golden Age of Astronomy, began to dawn. Orbiting observatories hundreds of thousands of times more sensitive than the observation stations of the past began to be put into use. As a result, astronomers are becoming able to probe the universe faster, more deeply and over a longer period of time than ever before. They are, in many cases, gathering data from radiations invisible to the human eye. And this Golden Age offers just one more avenue of promising discovery into regions of the unknown.

As Earth observers have begun to "see" into the hidden depths of the electromagnetic spectrum, they have begun recording the existence, shape, temperature and composition of such celestial objects as quasars, pulsars and supernovas that were previously hidden by the thick mantle of gas that blocked traditional viewing through visual light. One such new telescope observed clouds of interstellar gas and dust that appeared to be the actual birth of a group of new stars.

European Space Agency's X-ray astronomy satellite, put into orbit in 1983, was able to map sources of X-rays emitted in space with great accuracy and to help determine the temperature, density and chemical abundances of stellar gases. Four large American observatories will scan the far reaches of the electromagnetic spectrum in a quest to fathom the mysteries of the cosmos. They are the Space Infrared Telescope Facility, the Space Telescope, the Advanced X-ray Astrophysics Facility and the Gamma Ray Observatory. Each of these four telescopes will be tuned to a different part of the electromagnetic spectrum, picking up various wavelengths: in the infrared range, visual, ultraviolet, X-ray and gamma-ray.

These facilities are expected to enable scientists to see stars in the process of death, to see the gases of stars sucked into intense black holes whose mass is so compact and therefore so great and whose resulting gravity is so powerful that not even light can escape. The gamma-ray telescope, which detects the shortest wavelengths, will no doubt see some of the most violent events of all — the annihilation of matter and antimatter in exploding galaxies. The new telescopes are aimed at things that have until now defied accurate description or explanation. Gamma-ray bursters, for example, are points that occasionally emit explosive packets of gamma-ray energy, but they are yet to be linked with any known object in the universe.

In short, astronomers have begun to realize that explosions and dynamic change are far more common through the universe than they had thought. It is far more alive and, with the new equipment, far more observable. They now regard their work as comparable to Columbus's or Magellan's, believing that science is only just now beginning to learn the physical nature of the universe.

THE ULTIMATE MACHINE

Questions about the unknown could conceivably all come together — either in theory or in some pracitical expression of it — in one of the great scientific and technological issues of the twentieth century. If a computer can be taught to modify its own programs — and to a large extent it has already learned to do so — could not it be taught to reproduce? Indeed, the most profound impact of the computer could conceivably turn out to arise not from its power as an incredibily rapid problem solver but rather from its potential for self-replication — even mutation. In other words, given all that science now knows about the human brain and nervous system, given the help that this knowledge provided in the theoretical invention of the computer itself, given the remarkable advances in physical technology and biotechnology, couldn't computers be taught to procreate? Why couldn't they give birth to themselves? And if you had a self-replicating computer why couldn't you send it into space to do all this remote exploring for you while you sat at home and watched from your armchair? Or, why couldn't it keep modifying its offspring to answer any conceivable question asked of it? Why couldn't it know, or learn to know, the unknowable?

When he was a young student at Harvard in the 1950s, Jeremy Bernstein, a physicist and writer, once had a chance encounter in Harvard Yard with the great John von Neumann, the mathematician who developed the theory of stored programming. Bernstein, a bit awestruck, approached him and asked, "Profesor von Neumann, will the computer ever replace the human mathematician?"

"Sonny," came the reply, "don't worry about it."

Much has happened since then. But as Bernstein pointed out many years after his encounter with von Neumann, no computer, by itself, has yet ever created anything, let alone a clone of itself. This has been true, for the most part, even in science fiction. The sorcerer's real apprentice did turn out to be a self-replicating machine — a broom that couldn't be stopped from carrying water. When it started creating a flood, the junior apprentice tried to chop it up with an ax. But every time he did the pieces turned into still more hustling, water-carrying brooms. Of course, it may have been that the sorcerer didn't really create a broom that created other brooms. He may simply have intervened each time to do the job himself — a nice point that will have to be left to wiser philosophers. Generally, science-fiction writers have relied on some kind of intervention by man. For example, take Arthur C. Clarke, who insists that it has been science-fiction writers like himself who have provided the broad structure of ideas for all twentieth-century technology except for the microchip — science-fiction writers never think small. In his short story *Into the Comet*, a computer fails and the ship heads for destruction when a Japanese crewman intervenes and teaches the crew to make abacuses, and the ship is miraculously saved. Again, Man to the rescue.

Still, it must be said that, because the computer owes so much in its own creation to simulations or imitations of biological aspects of the human brain, it is not illogical to suppose that it could imitate other human capacities as well. Reproduction, for example. Herman Goldstein points out in *The Computer from*

Pascal to von Neumann that von Neumann took much of his inspiration from physiologists' mathematical models of the human nervous system. In a famous paper, "A Logical Calculus of the Ideas Imminent in Nervous Activity," Warren S. McCullouch and Walter Pitts described the brain as a collection of cells called neurons that were connected by wirelike fibers that could transmit electric impulses in the manner of power-company relay stations. The neuron may be so wired as to require a pair, for example, to be fired before the signal can be relayed on to a third. This is a logic circuit, as in "If all A is B, and all B is C, then all A is C."

McCullouch and Pitts believed, with others, that even the most complicated mathematical statements can be reduced to such simple logical propositions. "Anything that can be exhaustively and unambiguously described," they wrote, "anything that can be completely and unambigously put into words, is ipso facto, realizable by a suitable finite neural network." In other words, the networks of brain cells can carry out processes of mathematical logic.

Von Neumann said that, to reproduce itself, a machine needed raw materials, a programming "factory" in which to work, a kind of photocopying device to duplicate and preserve programming instructions for future generations and a supervisor.

Von Neumann designed a computing machine much like the human brain except that he used vacuum tubes to serve as neurons. The network of wired-together vacuum tubes performed like the neural network. Theoretically, von Neumann's machine could do anything McCullouch's and Pitts's model of the mind and nervous system could do.

Since the human brain can teach itself to do all manner of things, could not the computer be programmed at least to recognize when its parts were wearing out and to order replacements on its display screen, if not build and install the machine itself? Well, to a large extent it does this already. Even a relatively small and inexpensive personal computer recognizes when its disk drives are malfunctioning: "BDOS Error on B," it may say.

So aren't we on the way to the self-perpetuating machine? Could we not teach the machine — program it — to reproduce? John Kemeny, former president of Dartmouth College, addressed the question first by asking one of his own: "What do we mean by reproduction? If we mean the creation of an object like the original out of nothing, then no machine can reproduce, but neither can a human being. The characteristic feature of the reproduction of life is that the living organism can create a new organism like itself out of inert matter surrounding it. If we agree that machines are not alive, and if we insist that the creation of life is an essential feature of reproduction, then we have begged the question. A machine cannot reproduce. So we must reformulate the problem in a way that won't make machine reproduction logically impossible. We must omit the word 'living.' We shall ask that the machine create a new organism like itself out of simple parts contained in the environment."

Computers now are really on the verge of being self-replicating. Their programs can already modify and rewrite their own programs. These modified programs are then used to supervise the construction of more computers. The biological component remains missing, and human intervention is still necessary. But self-replicating machinery appears more and more possible. The field of computer-aided engineering is progressing rapidly in that direction. Perhaps the day will really come with the advent of the biochip to replace the silicon chip. Or, it may come with the development of parallel processing that will enable the computer to perform the giant leaps the human mind performs when it plucks some bit of information from the enormous storage bin of its past. Or, when computers can "see" and "speak" much better than they do now.

And if the computer will indeed learn to self-replicate, what then? Could we not send platoons of them to colonize the moon or Mars or some newly discovered planet, like Planet X, which some scientists believe may be lurking undiscovered on the edge of the solar system? Could not colonies of such machines accomplish all manner of good, or evil, right here on Earth? Freeman Dyson of the Institute for Advanced Study at Princeton said in a lecture in 1970 that a fully developed colony of self-replicating computers "must be as well coordinated as the cells of a bird. There must be automata with specialized functions corresponding to muscle, liver and nerve cell. There must be high-quality sense organs and a central battery of computers performing the functions of a brain,'' which may mutate and proliferate.

Though clearly fascinated by the subject, though clearly convinced that the future of propagating colonies of computers was in the realm of possibility, Jeremy Bernstein wrote about it in *The New York Times Magazine* with trepidation mixed with fascination. "For some reason," he said, "as admiring as I am of the logic of automation, I find the prospect chilling."

THE BIRTH OF A FUTURE

In science, indeed, in virtually all human endeavor, there is ultimately an exciting kind of vagueness and uncertainty about what we know and what we don't know. The creative processes of the truly great scientist, like the great painter or poet, produce knowledge or insights or new forms or simple new statements of truth that in some way may have been taking shape long before the thinker himself knew they were there. Robert Frost once said he was astounded by things he knew that he didn't know he knew. Of all developments in science, Stephen Toulmin, a professor of social thought and philosophy at the University of Chicago, writes in the journal *The Skeptical Inquirer*: "There is a point beyond which we cannot know for certain exactly what it is that we do know and we do not understand and exactly where a line should be drawn between phenomena that are as yet mysterious and happenings that are frankly incredible." The messages of the past are clear. Worry not about the known. Don't let it outwit you. Don't let it frighten you. But don't ignore it. In important ways, we already know the future, though like Frost, we don't yet know we know it. Or put another way, the future is the unknown, the unobserved, as yet unobservable present. Discoveries build on discoveries, knowledge on knowledge, surprise on surprise. Serendipity favors the prepared mind.

Twentieth-century science was built to a large degree on the discoveries of the nineteenth, despite its own profound skepticism about them — what member of the modern world of electronics, for example, holds any of the doubts now about James Maxwell's "scandalous" theory of electromagnetism that distinguished nineteenth-century scientists had? So, too, will the coming era be constructed on the present. The fascination is that in many ways we know that the twenty-first century will be profoundly different from the twentieth. In science, we have come to understand enough of the present and the past to sense the exciting nature of that difference. In a sense it's both humorous and encouraging that this excitement does go so far into the past.

Isn't it ironic that Pythagoras got into so much trouble that he was forced to flee ancient Magna Graecia because of his religious and philosphical teachings: that numbers constitute the true nature of things, that all philosophy resolved into the relation of numbers? This did, indeed, seem heretical. How could it be that the gods, that the complicated emotions of man, that intelligent verbal communication all resolved themselves into the relation of numbers? But leap 2,000 years forward. Leap to the quantum theory. Leap to the Grand Unification Theories. If it's true that all the physical forces of nature can be explained by a single theory, if commonality can be found among the vectors of those forces, if there is, indeed, a unity in all things physical, one seems compelled as Pythagoras was, or, if you will, as Einstein was, to become mystical about it. The point is not so much that theology can be reduced to numbers, that the gods were or are merely some cold and abstract set of calculations. Rather that clarity, symmetry, purity, neatness, unity can be achieved through physical theory and numbers. "Einstein was one of the great unifiers in the history of physics," as Timothy Ferris has written. "He had an unwavering faith in the unity of the universe, and he refused to be discouraged in his attempts to draw together widely disparate ideas." But even if unity is not an end in itself — though it may seem a better goal than most — then who is to say that it will not lead on to something else? This is the nature of science. This is the nature of exploration.

Whether the discovery will have anything to do with numbers, it must, indeed, have much to do with both persistence and imagination. Paul Dirac, one of the founders of quantum mechanics, once told an interviewer who asked where theoretical ideas come from: "You just have to try and imagine what the universe is like." That is what the creative minds of the twentieth century have been about: trying to image what the universe is like. They have succeeded to an astounding degree, and each time they succeed, they have created a remarkably clear image of the future as well. So in this sense, the future belongs to all those who are willing to take part in it now. The future is potentially everybody's. It belongs to those with the gift of inspired vision. It belongs to the philosopher, the poet, the dreamer, the painter, the statesman, the scientist. From the arguments and accounts put forth by the authors of this book, from the science and technology of the past and present they describe, the mists around the future begin to clear, for it has already been born in the creative, exciting mind.